Programming Persistent Memory

A Comprehensive Guide for Developers

Steve Scargall

Programming Persistent Memory: A Comprehensive Guide for Developers

Steve Scargall
Santa Clara, CA, USA

ISBN-13 (pbk): 978-1-4842-4931-4   ISBN-13 (electronic): 978-1-4842-4932-1
https://doi.org/10.1007/978-1-4842-4932-1

Managing Director, Apress Media LLC: Welmoed Spahr
Acquisitions Editor: Susan McDermott
Development Editor: Laura Berendson
Coordinating Editor: Jessica Vakili

Distributed to the book trade worldwide by Springer Science+Business Media New York, 233 Spring Street, 6th Floor, New York, NY 10013. Phone 1-800-SPRINGER, fax (201) 348-4505, e-mail orders-ny@springer-sbm.com, or visit www.springeronline.com. Apress Media, LLC is a California LLC and the sole member (owner) is Springer Science + Business Media Finance Inc (SSBM Finance Inc). SSBM Finance Inc is a **Delaware** corporation.

For information on translations, please e-mail rights@apress.com, or visit http://www.apress.com/rights-permissions.

Apress titles may be purchased in bulk for academic, corporate, or promotional use. eBook versions and licenses are also available for most titles. For more information, reference our Print and eBook Bulk Sales web page at http://www.apress.com/bulk-sales.

Any source code or other supplementary material referenced by the author in this book is available to readers on GitHub via the book's product page, located at www.apress.com/978-1-4842-4931-4. For more detailed information, please visit http://www.apress.com/source-code.

Printed on acid-free paper

Table of Contents

About the Author

Steve Scargall is a persistent memory software and cloud architect at Intel Corporation. As a technology evangelist, he supports the enabling and development effort to integrate persistent memory technology into software stacks, applications, and hardware architectures. This includes working with independent software vendors (ISVs) on both proprietary and open source development, original equipment manufacturers (OEMs), and cloud service providers (CSPs).

Steve holds a Bachelor of Science in computer science and cybernetics from the University of Reading, UK, where he studied neural networks, AI, and robotics. He has over 19 years' experience providing performance analysis on x86 architecture and SPARC for Solaris Kernel, ZFS, and UFS file system. He performed DTrace debugging in enterprise and cloud environments during his tenures at Sun Microsystems and Oracle.

About the Technical Reviewer

Andy Rudoff is a principal engineer at Intel Corporation, focusing on non-volatile memory programming. He is a contributor to the SNIA NVM Programming Technical Work Group. His more than 30 years' industry experience includes design and development work in operating systems, file systems, networking, and fault management at companies large and small, including Sun Microsystems and VMware. Andy has taught various operating systems classes over the years and is a coauthor of the popular *UNIX Network Programming* textbook.

About the Contributors

Piotr Balcer is a software engineer at Intel Corporation with many years' experience working on storage-related technologies. He holds a Bachelor of Science in engineering from the Gdańsk University of Technology, Poland, where he studied system software engineering. Piotr has been working on the software ecosystem for next-generation persistent memory since 2014.

Eduardo Berrocal joined Intel Corporation as a cloud software engineer in 2017 after receiving his PhD in computer science from the Illinois Institute of Technology. His doctoral research focused on data analytics and fault tolerance for high-performance computing. Past experience includes working as an intern at Bell Labs (Nokia), a research aid at Argonne National Laboratory, a scientific programmer and web developer at the University of Chicago, and an intern in the CESVIMA laboratory in Spain.

Adam Borowski is a software engineer at Intel Corporation, hailing from the University of Warsaw, Poland. He is a Debian developer and.has made many open source contributions over the past two decades. Adam is currently working on persistent memory stacks, both on upstream code and integrating it with downstream distributions.

Igor Chorazewicz is a software engineer at Intel Corporation. His main focus is on persistent memory data structures and enabling C++ applications for persistent memory. Igor holds a Bachelor of Science in engineering from the Gdańsk University of Technology, Poland.

Adam Czapski is a technical writer at Intel Corporation. He writes technical documentation in the Data Center Group and is currently working in the persistent memory department. Adam holds a Bachelor of Arts in English philology and a master's degree in natural language processing from the Gdańsk University of Technology, Poland.

Steve Dohrmann is a software engineer at Intel Corporation. He has worked on a variety of projects over the past 20 years, including media frameworks, mobile agent software, secure collaboration software, and parallel programming language implementation. He is currently working on enabling the use of persistent memory in Java*.

Chet Douglas is a principal software engineer at Intel Corporation and focuses on cloud software architecture along with operating system and OEM enabling of non-volatile memory technologies. He has over 14 years' experience working on various enterprise and client programs and 28 years of total storage experience. Chet has worked in all aspects of storage, including storage controller hardware design, SCSI disk/tape/CD writer firmware architecture, storage management software architecture, Microsoft Windows* and Linux kernel-mode drivers, enterprise hardware RAID, and client/workstation software RAID. He holds seven storage-related hardware and software patents and has a dual Bachelor of Science in electrical engineering and computer engineering from Clarkson University, New York.

Ken Gibson is the director of persistent memory software architecture within Intel Corporation's Data Center Group. Since 2012, Ken and his team have been working with Intel's server and software partners to create the open persistent memory programming model.

Tomasz Gromadzki is a software architect in Intel Corporation's Non-Volatile Memory Solutions Group. His focus is on remote persistent memory access, which includes proper integration of persistent memory with other (networking) technologies as well as optimal persistent memory replication procedures and algorithms.

Kishor Kharbas is a software engineer on the Java runtime engineering team at Intel Corporation. For the past eight years, he has been working to optimize Oracle's OpenJDK on Intel platforms. This involves Java garbage collection and compiler back-end optimization.

Jackson Marusarz is a senior technical consulting engineer (TCE) in Intel Corporation's Compute Performance and Developer Products Division. As the lead TCE for Intel VTune Profiler, his main focus is on software performance analysis and tuning for both serial and multithreaded applications. Jackson's time is split between determining how to analyze and tune software and creating tools that help others do the same.

Jan Michalski is a software engineer in Intel Corporation's Non-Volatile Memory Solutions Group. His focus is on remote persistent memory access, which includes proper integration of persistent memory with other technologies, as well as looking for optimal persistent memory replication procedures and algorithms. He holds a master's degree in computer engineering from the Gdańsk University of Technology, Poland, where he studied system software engineering.

Nicholas Moulin is a cloud software architect at Intel Corporation. Since joining Intel in 2012, he has focused on enabling and developing persistent memory software for operating systems and platform firmware and managing persistent memory hardware. Nicholas is currently working with industry partners to define and improve RAS features relevant to the persistent memory programming model.

Szymon Romik is a software engineer at Intel Corporation and is currently focused on persistent memory programming. He previously worked as a lead software engineer on 5G technologies at Ericsson. Szymon holds a master's degree in mathematics from Jagiellonian University, Poland.

Jakub Schmiegel is a software architect in Intel Corporation's Non-Volatile Memory Solutions Group where he has been focused on enabling existing applications to persistent memory and analyzing their performance for more than four years. Jakub holds a master's degree in computer science from the Gdańsk University of Technology, Poland.

Kevin Shalkowsky is a Telly Award–winning creative director, graphic designer, and animator with more than a decade of experience. While his contributions are in technology today, Kevin has spent time in broadcast journalism and selling numerous products through 30-minute late-night infomercials. He resides in Oregon with his wife and son. From time to time, you can find Kevin lost in the woods, lost in a parking lot, or lost in his design process – but this somehow got him to where he is today, and he wouldn't have it any other way.

Vineet Singh is a memory and storage tools software engineer at Intel Corporation. He develops techniques to help developers adapt to the latest memory technologies. Vineet holds a PhD in philosophy from the University of California and has a Bachelor of Technology degree from the Indian Institute of Information Technology, Design, and Manufacturing in Jabalpur.

Pawel Skowron is a software engineering manager at Intel Corporation with 20 years' experience in the software industry. Pawel has worked in various roles related to the whole-software development life cycle. His software engineering background lies in the areas of embedded systems, database systems, and applications. For the past few years, Pawel has led the development and validation of the Persistent Memory Development Kit (`https://github.com/pmem/pmdk`).

Usha Upadhyayula has been with Intel Corporation for 20 years serving in many different roles. Usha holds a master's degree in computer science from the University of South Carolina, and she spent the first few years at Intel developing user-level applications in C and C++. She later moved to customer-enabling roles for Intel media processors and support for Intel RAID software. Usha is currently part of the Data Center Group where she is focused on enabling cloud service providers to fully utilize and accelerate the adoption of Intel persistent memory products.

Sergey Vinogradov is a senior software development engineer at Intel Corporation where he spent more than seven years working on performance profiling tools and threading runtime libraries. During the past four years, Sergey has been working on C++ programming models and performance profiling methodologies for persistent memory.

Acknowledgments

First and foremost, I would like to thank Ken Gibson for masterminding this book idea and for gifting me the pleasure of writing and managing it. Your support, guidance, and contributions have been instrumental in delivering a high-quality product.

If the Vulcan mind-meld or *The Matrix* Headjack were possible, I could have cloned Andy Rudoff's mind and allowed him to work on his daily activities. Instead, Andy's infinite knowledge of persistent memory had to be tapped through good old verbal communication and e-mail. I sincerely thank you for devoting so much time to me and this project. The results read for themselves.

Debbie Graham was instrumental in helping me manage this colossal project. Her dedication and support helped drive the project to an on-time completion.

To my friends and colleagues at Intel who contributed content, supported discussions, helped with decision-making, and reviewed drafts during the book-writing process. These are the real heroes. Without your heavily invested time and support, this book would have taken considerably longer to complete. It is a much better product as a result of the collaborative effort. A huge thanks to all of you.

I'd like to express my sincerest gratitude and appreciation to the people at Apress, without whom this book could not have been published. From the initial contact and outline discussions through the entire publishing process to this final polished product, the Apress team delivered continuous support and assistance. Many thanks to Susan, Jessica, and Rita. It was a real pleasure working with you.

Preface

About This Book

Persistent memory is often referred to as non-volatile memory (NVM) or storage class memory (SCM). In this book, we purposefully use *persistent memory* as an all-encompassing term to represent all the current and future memory technologies that fall under this umbrella. This book introduces the persistent memory technology and provides answers to key questions. For software developers, those questions include: What is persistent memory? How do I use it? What APIs and libraries are available? What benefits can it provide for my application? What new programming methods do I need to learn? How do I design applications to use persistent memory? Where can I find information, documentation, and help?

System and cloud architects will be provided with answers to questions such as: What is persistent memory? How does it work? How is it different than DRAM or SSD/NVMe storage devices? What are the hardware and operating system requirements? What applications need or could benefit from persistent memory? Can my existing applications use persistent memory without being modified?

Persistent memory is not a plug-and-play technology for software applications. Although it may look and feel like traditional DRAM memory, applications need to be modified to fully utilize the persistence feature of persistent memory. That is not to say that applications cannot run unmodified on systems with persistent memory installed, they can, but they will not see the full potential of what persistent memory offers without code modification.

Thankfully, server and operating system vendors collaborated very early in the design phase and already have products available on the market. Linux and Microsoft Windows already provide native support for persistent memory technologies. Many popular virtualization technologies also support persistent memory.

For ISVs and the developer community at large, the journey is just beginning. Some software has already been modified and is available on the market. However, it will take time for the enterprise and cloud computing industries to adopt and make the hardware available to the general marketplace. ISVs and software developers need time to understand what changes to existing applications are required and implement them.

To make the required development work easier, Intel developed and open sourced the Persistent Memory Development Kit (PMDK) available from `https://pmem.io/pmdk/`. We introduce the PMDK in more detail in Chapter 5 and walk through most of the available libraries in subsequent chapters. Each chapter provides an in-depth guide so developers can understand what library or libraries to use. PMDK is a set of open source libraries and tools based on the Storage Networking Industry Association (SNIA) NVM programming model designed and implemented by over 50 industry partners. The latest NVM programming model document can be found at `https://www.snia.org/tech_activities/standards/curr_standards/npm`. The model describes how software can utilize persistent memory features and enables designers to develop APIs that take advantage of NVM features and performance.

Available for both Linux and Windows, PMDK facilitates persistent memory programming adoption with higher-level language support. C and C++ support is fully validated. Support for other languages such as Java and Python is work in progress at the time this book was written. Other languages are expected to also adopt the programming model and provide native persistent memory APIs for developers. The PMDK development team welcomes and encourages new contributions to core code, new language bindings, or new storage engines for the persistent memory key-value store called `pmemkv`.

This book assumes no prior knowledge of persistent memory hardware devices or software development. The book layout allows you to freely navigate the content in the order you want. It is not required to read all chapters in order, though we do build upon concepts and knowledge described in previous chapters. In such cases, we make backward and forward references to relevant chapters and sections so you can learn or refresh your memory.

Book Structure

This book has 19 chapters, each one focusing on a different topic. The book has three main sections. Chapters 1-4 provide an introduction to persistent memory architecture, hardware, and operating system support. Chapters 5-16 allow developers to understand the PMDK libraries and how to use them in applications. Finally, Chapters 17-19 provide information on advanced topics such as RAS and replication of data using RDMA.

- Chapter 1. Introduction to Persistent Memory – *Introduces persistent memory and dips our toes in the water with a simple persistent key-value store example using libpmemkv.*

- Chapter 2. Persistent Memory Architecture – *Describes the persistent memory architecture and focuses on the hardware requirements developers should know.*

- Chapter 3. Operating System Support for Persistent Memory – *Provides information relating to operating system changes, new features, and how persistent memory is seen by the OS.*

- Chapter 4. Fundamental Concepts of Persistent Memory Programming – *Builds on the first three chapters and describes the fundamental concepts of persistent memory programming.*

- Chapter 5. Introducing the Persistent Memory Development Kit (PMDK) – *Introduces the Persistent Memory Development Kit (PMDK), a suite of libraries to assist software developers.*

- Chapter 6. libpmem: Low-Level Persistent Memory Support – *Describes and shows how to use libpmem from the PMDK, a low-level library providing persistent memory support.*

- Chapter 7. libpmemobj: A Native Transactional Object Store – *Provides information and examples using libpmemobj, a C native object store library from the PMDK.*

- Chapter 8. libpmemobj-cpp: The Adaptable Language - C++ and Persistent Memory – *Demonstrates the C++ libpmemobj-cpp object store from the PMDK, built using C++ headers on top of libpmemobj.*

- Chapter 9. pmemkv: A Persistent In-Memory Key-Value Store – *Expands upon the introduction to libpmemkv from Chapter 1 with a more in-depth discussion using examples.*

- Chapter 10. Volatile Use of Persistent Memory – *This chapter is for those who want to take advantage of persistent memory but do not require data to be stored persistently. libmemkind is a user-extensible heap manager built on top of jemalloc which enables control of memory characteristics and a partitioning of the heap*

*between different kinds of memory, including persistent memory.
libvmemcache is an embeddable and lightweight in-memory caching
solution. It is designed to fully take advantage of large-capacity
memory, such as persistent memory with DAX, through memory
mapping in an efficient and scalable way.*

- Chapter 11. Designing Data Structures for Persistent Memory – *Provides a wealth of information for designing data structures for persistent memory.*

- Chapter 12. Debugging Persistent Memory Applications – *Introduces tools and walks through several examples for how software developers can debug persistent memory–enabled applications.*

- Chapter 13. Enabling Persistence using a Real-World Application – *Discusses how a real-world application was modified to enable persistent memory features.*

- Chapter 14. Concurrency and Persistent Memory – *Describes how concurrency in applications should be implemented for use with persistent memory.*

- Chapter 15. Profiling and Performance – *Teaches performance concepts and demonstrates how to use the Intel VTune suite of tools to profile systems and applications before and after code changes are made.*

- Chapter 16. PMDK Internals: Important Algorithms and Data Structures – *Takes us on a deep dive of the PMDK design, architecture, algorithms, and memory allocator implementation.*

- Chapter 17. Reliability, Availability, and Serviceability (RAS) – *Describes the implementation of reliability, availability, and serviceability (RAS) with the hardware and operating system layers.*

- Chapter 18. Remote Persistent Memory – *Discusses how applications can scale out across multiple systems using local and remote persistent memory.*

- Chapter 19. Advanced Topics – *Describes things such as NUMA, using software volume managers, and the mmap() MAP_SYNC flag.*

The Appendixes have separate procedures for installing the PMDK and utilities required for managing persistent memory. We also included an update for Java and the future of the RDMA protocols. All of this content is considered temporal, so we did not want to include it in the main body of the book.

Intended Audience

This book has been written for experienced application developers in mind. We intend the content to be useful to a wider readership such as system administrators and architects, students, lecturers, and academic research fellows to name but a few. System designers, kernel developers, and anyone with a vested or passing interest in this emerging technology will find something useful within this book.

Every reader will learn what persistent memory is, how it works, and how operating systems and applications can utilize it. Provisioning and managing persistent memory are vendor specific, so we include some resources in the Appendix sections to avoid overcomplicating the main chapter content.

Application developers will learn, by example, how to integrate persistent memory in to existing or new applications. We use examples extensively throughout this book using a variety of libraries available within the Persistent Memory Development Kit (PMDK). Example code is provided in a variety of programming languages such as C, C++, JavaScript, and others. We want developers to feel comfortable using these libraries in their own projects. The book provides extensive links to resources where you can find help and information.

System administrators and architects of Cloud, high-performance computing, and enterprise environments can use most of the content of this book to understand persistent memory features and benefits to support applications and developers. Imagine being able to deploy more virtual machines per physical server or provide applications with this new memory/storage tier such that they can keep more data closer to the CPU or restart in a fraction of the time they could before while keeping a warm cache of data.

Students, lecturers, and academic research fellows will also benefit from many chapters within this book. Computer science classes can learn about the hardware, operating system features, and programming techniques. Lecturers are free use the content in student classes or to form the basis of research projects such as new persistent memory file systems, algorithms, or caching implementations.

We introduce tools that profile the server and applications to better understand CPU, memory, and disk IO access patterns. Using this knowledge, we show how applications can be modified to take full advantage of persistence using the Persistent Memory Development Kit (PMDK).

A Future Reference

The book content has been written to provide value for many years. Industry specification such as ACPI, UEFI, and the SNIA non-volatile programming model will, unless otherwise stated by the specification, remain backward compatible as new versions are released. If new form factors are introduced, the approach to programming remains the same. We do not limit ourselves to one specific persistent memory vendor or implementation. In places where it is necessary to describe vendor-specific features or implementations, we specifically call this out as it may change between vendors or between product generations. We encourage you to read the vendor documentation for the persistent memory product to learn more.

Developers using the Persistent Memory Development Kit (PMDK) will retain a stable API interface. PMDK will deliver new features and performance improvements with each major release. It will evolve with new persistent memory products, CPU instructions, platform designs, industry specifications, and operating system feature support.

Source Code Examples

Concepts and source code samples within this book adhere to the vendor neutral SNIA non-volatile memory programming model. SNIA which is the Storage Networking Industry Association is a non-profit global organization dedicated to developing standards and education programs to advance storage and information technology. The model was designed, developed, and is maintained by the SNIA NVM Technical Working Group (TWG) which includes many leading operating system, hardware, and server vendors. You can join this group or find information at `https://www.snia.org/forums/sssi/nvmp`.

The code examples provided with this book have been tested and validated using Intel Optane DC persistent memory. Since the PMDK is vendor neutral, they will also work on NVDIMM-N devices. PMDK will support any future persistent memory product that enters the market.

The code examples used throughout this book are current at the time of publication. All code examples have been validated and tested to ensure they compile and execute without error. For brevity, some of the examples in this book use assert() statements to indicate unexpected errors. Any production code would likely replace these with the appropriate error handling actions which would include friendlier error messages and appropriate error recovery actions. Additionally, some of the code examples use different mount points to represent persistent memory aware file systems, for example "/daxfs", "/pmemfs", and "/mnt/pmemfs". This demonstrates persistent memory file systems can be mounted and named appropriately for the application, just like regular block-based file systems. Source code is from the repository that accompanies this book – `https://github.com/Apress/programming-persistent-memory`.

Since this is a rapidly evolving technology, the software and APIs references throughout this book may change over time. While every effort is made to be backward compatible, sometimes software must evolve and invalidate previous versions. For this reason, it is therefore expected that some of the code samples may not compile on newer hardware or operating systems and may need to be changed accordingly.

Book Conventions

This book uses several conventions to draw your attention to specific pieces of information. The convention used depends on the type of information displayed.

Computer Commands

Commands, programming library, and API function references may be presented in line with the paragraph text using a monospaced font. For example:

To illustrate how persistent memory is used, let's start with a sample program demonstrating the key-value store provided by a library called `libpmemkv`.

Computer Terminal Output

Computer terminal output is usually taken directly from a computer terminal presented in a monospaced font such as the following example demonstrating cloning the Persistent Memory Development Kit (PMDK) from the GitHub project:

```
$ git clone https://github.com/pmem/pmdk
Cloning into 'pmdk'...
remote: Enumerating objects: 12, done.
remote: Counting objects: 100% (12/12), done.
remote: Compressing objects: 100% (10/10), done.
remote: Total 100169 (delta 2), reused 7 (delta 2), pack-reused 100157
Receiving objects: 100% (100169/100169), 34.71 MiB | 4.85 MiB/s, done.
Resolving deltas: 100% (83447/83447), done.
```

Source Code

Source code examples taken from the accompanying GitHub repository are shown with relevant line numbers in a monospaced font. Below each code listing is a reference to the line number or line number range with a brief description. Code comments use language native styling. Most languages use the same syntax. Single line comments will use // and block/multiline comments should use /*..*/. An example is shown in Listing 1.

Listing 1. A sample program using libpmemkv

```
37  #include <iostream>
38  #include "libpmemkv.h"
39
40  using namespace pmemkv;
41
42  /*
43   * kvprint -- print a single key-value pair
44   */
45  void kvprint(const string& k, const string& v) {
46      std::cout << "key: " << k << ", value: " << v << "\n";
47  }
```

- Line 45: Here we define a small helper routine, `kvprint()`, which prints a key-value pair when called.

Notes

We use a standard format for notes, cautions, and tips when we want to direct your attention to an important point, for example.

Note   Notes are tips, shortcuts, or alternative approaches to the current discussion topic. Ignoring a note should have no negative consequences, but you might miss out on a nugget of information that makes your life easier.

CHAPTER 1

Introduction to Persistent Memory Programming

This book describes programming techniques for writing applications that use persistent memory. It is written for experienced software developers, but we assume no previous experience using persistent memory. We provide many code examples in a variety of programming languages. Most programmers will understand these examples, even if they have not previously used the specific language.

Note All code examples are available on a GitHub repository (`https://github.com/Apress/programming-persistent-memory`), along with instructions for building and running it.

Additional documentation for persistent memory, example programs, tutorials, and details on the Persistent Memory Development Kit (PMDK), which is used heavily in this book, can be found on `http://pmem.io`.

The persistent memory products on the market can be used in various ways, and many of these ways are transparent to applications. For example, all persistent memory products we encountered support the storage interfaces and standard file API's just like any solid-state disk (SSD). Accessing data on an SSD is simple and well-understood, so we consider these use cases outside the scope of this book. Instead, we concentrate on memory-style access, where applications manage byte-addressable data structures that reside in persistent memory. Some use cases we describe are *volatile*, using the persistent memory only for its capacity and ignoring the fact it is persistent. However, most of this book is dedicated to the *persistent* use cases, where data structures placed in persistent memory are expected to survive crashes and power failures, and the techniques described in this book keep those data structures consistent across those events.

1

© The Author(s) 2020
S. Scargall, *Programming Persistent Memory*, https://doi.org/10.1007/978-1-4842-4932-1_1

A High-Level Example Program

To illustrate how persistent memory is used, we start with a sample program demonstrating the key-value store provided by a library called libpmemkv. Listing 1-1 shows a full C++ program that stores three key-value pairs in persistent memory and then iterates through the key-value store, printing all the pairs. This example may seem trivial, but there are several interesting components at work here. Descriptions below the listing show what the program does.

Listing 1-1. A sample program using libpmemkv

```
37   #include <iostream>
38   #include <cassert>
39   #include <libpmemkv.hpp>
40
41   using namespace pmem::kv;
42   using std::cerr;
43   using std::cout;
44   using std::endl;
45   using std::string;
46
47   /*
48    * for this example, create a 1 Gig file
49    * called "/daxfs/kvfile"
50    */
51   auto PATH = "/daxfs/kvfile";
52   const uint64_t SIZE = 1024 * 1024 * 1024;
53
54   /*
55    * kvprint -- print a single key-value pair
56    */
57   int kvprint(string_view k, string_view v) {
58       cout << "key: "    << k.data() <<
59           " value: " << v.data() << endl;
60       return 0;
61   }
62
```

```
63  int main() {
64      // start by creating the db object
65      db *kv = new db();
66      assert(kv != nullptr);
67
68      // create the config information for
69      // libpmemkv's open method
70      config cfg;
71
72      if (cfg.put_string("path", PATH) != status::OK) {
73          cerr << pmemkv_errormsg() << endl;
74          exit(1);
75      }
76      if (cfg.put_uint64("force_create", 1) != status::OK) {
77          cerr << pmemkv_errormsg() << endl;
78          exit(1);
79      }
80      if (cfg.put_uint64("size", SIZE) != status::OK) {
81          cerr << pmemkv_errormsg() << endl;
82          exit(1);
83      }
84
85
86      // open the key-value store, using the cmap engine
87      if (kv->open("cmap", std::move(cfg)) != status::OK) {
88          cerr << db::errormsg() << endl;
89          exit(1);
90      }
91
92      // add some keys and values
93      if (kv->put("key1", "value1") != status::OK) {
94          cerr << db::errormsg() << endl;
95          exit(1);
96      }
```

```
 97      if (kv->put("key2", "value2") != status::OK) {
 98          cerr << db::errormsg() << endl;
 99          exit(1);
100      }
101      if (kv->put("key3", "value3") != status::OK) {
102          cerr << db::errormsg() << endl;
103          exit(1);
104      }
105
106      // iterate through the key-value store, printing them
107      kv->get_all(kvprint);
108
109      // stop the pmemkv engine
110      delete kv;
111
112      exit(0);
113 }
```

- Line 57: We define a small helper routine, kvprint(), which prints a key-value pair when called.

- Line 63: This is the first line of main() which is where every C++ program begins execution. We start by instantiating a key-value engine using the engine name "cmap". We discuss other engine types in Chapter 9.

- Line 70: The cmap engine takes config parameters from a config structure. The parameter "path" is configured to "/daxfs/kvfile", which is the path to a persistent memory file on a DAX file system; the parameter "size" is set to SIZE. Chapter 3 describes how to create and mount DAX file systems.

- Line 93: We add several key-value pairs to the store. The trademark of a key-value store is the use of simple operations like put() and get(); we only show put() in this example.

- Line 107: Using the get_all() method, we iterate through the entire key-value store, printing each pair when get_all() calls our kvprint() routine.

What's Different?

A wide variety of key-value libraries are available in practically every programming language. The persistent memory example in Listing 1-1 is different because the key-value store itself resides in persistent memory. For comparison, Figure 1-1 shows how a key-value store using traditional storage is laid out.

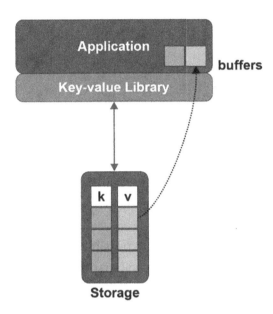

Figure 1-1. *A key-value store on traditional storage*

When the application in Figure 1-1 wants to fetch a value from the key-value store, a buffer must be allocated in memory to hold the result. This is because the values are kept on block storage, which cannot be addressed directly by the application. The only way to access a value is to bring it into memory, and the only way to do that is to read full blocks from the storage device, which can only be accessed via block I/O. Now consider Figure 1-2, where the key-value store resides in persistent memory like our sample code.

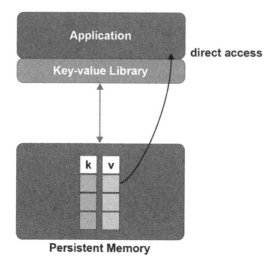

Figure 1-2. *A key-value store in persistent memory*

With the persistent memory key-value store, values are accessed by the application directly, without the need to first allocate buffers in memory. The kvprint() routine in Listing 1-1 will be called with references to the actual keys and values, directly where they live in persistence – something that is not possible with traditional storage. In fact, even the data structures used by the key-value store library to organize its data are accessed directly. When a storage-based key-value store library needs to make a small update, for example, 64 bytes, it must read the block of storage containing those 64 bytes into a memory buffer, update the 64 bytes, and then write out the entire block to make it persistent. That is because storage accesses can only happen using block I/O, typically 4K bytes at a time, so the task to update 64 bytes requires reading 4K and then writing 4K. But with persistent memory, the same example of changing 64 bytes would only write the 64 bytes directly to persistence.

The Performance Difference

Moving a data structure from storage to persistent memory does not just mean smaller I/O sizes are supported; there is a fundamental performance difference. To illustrate this, Figure 1-3 shows a hierarchy of latency among the different types of media where data can reside at any given time in a program.

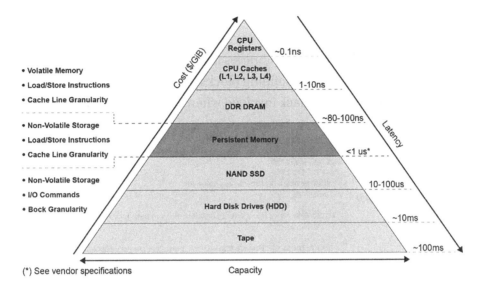

Figure 1-3. *The memory/storage hierarchy pyramid with estimated latencies*

As the pyramid shows, persistent memory provides latencies similar to memory, measured in nanoseconds, while providing persistency. Block storage provides persistency with latencies starting in the microseconds and increasing from there, depending on the technology. Persistent memory is unique in its ability to act like both memory and storage at the same time.

Program Complexity

Perhaps the most important point of our example is that the programmer still uses the familiar get/put interfaces normally associated with key-value stores. The fact that the data structures are in persistent memory is abstracted away by the high-level API provided by libpmemkv. This principle of using the highest level of abstraction possible, as long as it meets the application's needs, will be a recurring theme throughout this book. We start by introducing very high-level APIs; later chapters delve into the lower-level details for programmers who need them. At the lowest level, programming directly to raw persistent memory requires detailed knowledge of things like hardware atomicity, cache flushing, and transactions. High-level libraries like libpmemkv abstract away all that complexity and provide much simpler, less error-prone interfaces.

How Does libpmemkv Work?

All the complexity hidden by high-level libraries like libpmemkv are described more fully in later chapters, but let's look at the building blocks used to construct a library like this. Figure 1-4 shows the full software stack involved when an application uses libpmemkv.

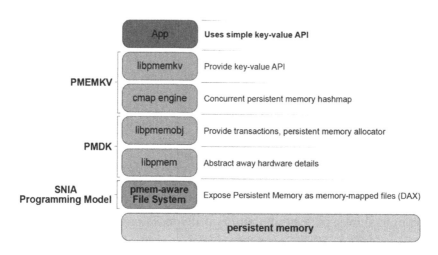

Figure 1-4. *The software stack when using* libpmemkv

Starting from the bottom of Figure 1-4 and working upward are these components:

- The *persistent memory* hardware, typically connected to the system memory bus and accessed using common memory load/store operations.

- A *pmem-aware file system*, which is a kernel module that exposes persistent memory to applications as files. Those files can be memory mapped to give applications direct access (abbreviated as DAX). This method of exposing persistent memory was published by SNIA (Storage Networking Industry Association) and is described in detail in Chapter 3.

- The libpmem library is part of the PMDK. This library abstracts away some of the low-level hardware details like cache flushing instructions.

- The libpmemobj library is a full-featured transaction and allocation library for persistent memory. (Chapters 7 and 8 describe libpmemobj and its C++ cousin in more detail.) If you cannot find data structures that meet your needs, you will most likely have to implement what you need using this library, as described in Chapter 11.

- The *cmap* engine, a concurrent hash map optimized for persistent memory.

- The libpmemkv library, providing the API demonstrated in Listing 1-1.

- And finally, the application that uses the API provided by libpmemkv.

Although there is quite a stack of components in use here, it does not mean there is necessarily a large amount of code that runs for each operation. Some components are only used during the initial setup. For example, the pmem-aware file system is used to find the persistent memory file and perform permission checks; it is out of the application's data path after that. The PMDK libraries are designed to leverage the direct access allowed by persistent memory as much as possible.

What's Next?

Chapters 1 through 3 provide the essential background that programmers need to know to start persistent memory programming. The stage is now set with a simple example; the next two chapters provide details about persistent memory at the hardware and operating system levels. The later and more advanced chapters provide much more detail for those interested.

Because the immediate goal is to get you programming quickly, we recommend reading Chapters 2 and 3 to gain the essential background and then dive into Chapter 4 where we start to show more detailed persistent memory programming examples.

Summary

This chapter shows how high-level APIs like libpmemkv can be used for persistent memory programming, hiding complex details of persistent memory from the application developer. Using persistent memory can allow finer-grained access and higher performance than block-based storage. We recommend using the highest-level, simplest APIs possible and only introducing the complexity of lower-level persistent memory programming as necessary.

CHAPTER 2

Persistent Memory Architecture

This chapter provides an overview of the persistent memory architecture while focusing on the hardware to emphasize requirements and decisions that developers need to know.

Applications that are designed to recognize the presence of persistent memory in a system can run much faster than using other storage devices because data does not have to transfer back and forth between the CPU and slower storage devices. Because applications that only use persistent memory may be slower than dynamic random-access memory (DRAM), they should decide what data resides in DRAM, persistent memory, and storage.

The capacity of persistent memory is expected to be many times larger than DRAM; thus, the volume of data that applications can potentially store and process in place is also much larger. This significantly reduces the number of disk I/Os, which improves performance and reduces wear on the storage media.

On systems without persistent memory, large datasets that cannot fit into DRAM must be processed in segments or streamed. This introduces processing delays as the application stalls waiting for data to be paged from disk or streamed from the network.

If the working dataset size fits within the capacity of persistent memory and DRAM, applications can perform in-memory processing without needing to checkpoint or page data to or from storage. This significantly improves performance.

© The Author(s) 2020
S. Scargall, *Programming Persistent Memory*, https://doi.org/10.1007/978-1-4842-4932-1_2

Persistent Memory Characteristics

As with every new technology, there are always new things to consider. Persistent memory is no exception. Consider these characteristics when architecting and developing solutions:

- Performance (throughput, latency, and bandwidth) of persistent memory is much better than NAND but potentially slower than DRAM.

- Persistent memory is durable unlike DRAM. Its endurance is usually orders of magnitude better than NAND and should exceed the lifetime of the server without wearing out.

- Persistent memory module capacities can be much larger than DRAM DIMMs and can coexist on the same memory channels.

- Persistent memory-enabled applications can update data in place without needing to serialize/deserialize the data.

- Persistent memory is byte addressable like memory. Applications can update only the data needed without any read-modify-write overhead.

- Data is CPU cache coherent.

- Persistent memory provides direct memory access (DMA) and remote DMA (RDMA) operations.

- Data written to persistent memory is not lost when power is removed.

- After permission checks are completed, data located on persistent memory is directly accessible from user space. No kernel code, file system page caches, or interrupts are in the data path.

- Data on persistent memory is instantly available, that is:

  - Data is available as soon as power is applied to the system.

  - Applications do not need to spend time warming up caches. They can access the data immediately upon memory mapping it.

  - Data residing on persistent memory has no DRAM footprint unless the application copies data to DRAM for faster access.

- Data written to persistent memory modules is local to the system. Applications are responsible for replicating data across systems.

Platform Support for Persistent Memory

Platform vendors such as Intel, AMD, ARM, and others will decide how persistent memory should be implemented at the lowest hardware levels. We try to provide a vendor-agnostic perspective and only occasionally call out platform-specific details.

For systems with persistent memory, failure atomicity guarantees that systems can always recover to a consistent state following a power or system failure. Failure atomicity for applications can be achieved using logging, flushing, and memory store barriers that order such operations. Logging, either undo or redo, ensures atomicity when a failure interrupts the last atomic operation from completion. Cache flushing ensures that data held within volatile caches reach the persistence domain so it will not be lost if a sudden failure occurs. Memory store barriers, such as an SFENCE operation on the x86 architecture, help prevent potential reordering in the memory hierarchy, as caches and memory controllers may reorder memory operations. For example, a barrier ensures that the undo log copy of the data gets persisted onto the persistent memory before the actual data is modified in place. This guarantees that the last atomic operation can be rolled back should a failure occur. However, it is nontrivial to add such failure atomicity in user applications with low-level operations such as write logging, cache flushing, and barriers. The Persistent Memory Development Kit (PMDK) was developed to isolate developers from having to re-implement the hardware intricacies.

Failure atomicity should be a familiar concept, since most file systems implement and perform journaling and flushing of their metadata to storage devices.

Cache Hierarchy

We use load and store operations to read and write to persistent memory rather than using block-based I/O to read and write to traditional storage. We suggest reading the CPU architecture documentation for an in-depth description because each successive CPU generation may introduce new features, methods, and optimizations.

Using the Intel architecture as an example, a CPU cache typically has three distinct levels: L1, L2, and L3. The hierarchy makes references to the distance from the CPU core, its speed, and size of the cache. The L1 cache is closest to the CPU. It is extremely fast but very small. L2 and L3 caches are increasingly larger in capacity, but they are relatively slower. Figure 2-1 shows a typical CPU microarchitecture with three levels of CPU cache and a memory controller with three memory channels. Each memory channel has a single DRAM and persistent memory attached. On platforms where the CPU caches are not contained within the power-fail protected domain, any modified data within the CPU caches that has not been flushed to persistent memory will be lost when the system loses power or crashes. Platforms that do include CPU caches in the power-fail protected domain will ensure modified data within the CPU caches are flushed to the persistent memory should the system crash or loses power. We describe these requirements and features in the upcoming "Power-Fail Protected Domains" section.

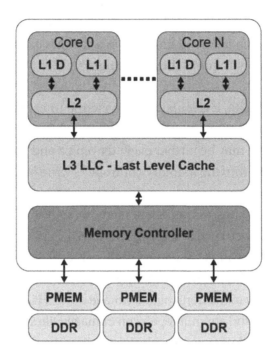

Figure 2-1. *CPU cache and memory hierarchy*

The L1 (Level 1) cache is the fastest memory in a computer system. In terms of access priority, the L1 cache has the data the CPU is most likely to need while completing a specific task. The L1 cache is also usually split two ways, into the instruction cache (L1 I) and the data cache (L1 D). The instruction cache deals with the information about the operation that the CPU has to perform, while the data cache holds the data on which the operation is to be performed.

The L2 (Level 2) cache has a larger capacity than the L1 cache, but it is slower. L2 cache holds data that is likely to be accessed by the CPU next. In most modern CPUs, the L1 and L2 caches are present on the CPU cores themselves, with each core getting dedicated caches.

The L3 (Level 3) cache is the largest cache memory, but it is also the slowest of the three. It is also a commonly shared resource among all the cores on the CPU and may be internally partitioned to allow each core to have dedicated L3 resources.

Data read from DRAM or persistent memory is transferred through the memory controller into the L3 cache, then propagated into the L2 cache, and finally the L1 cache where the CPU core consumes it. When the processor is looking for data to carry out an operation, it first tries to find it into the L1 cache. If the CPU can find it, the condition is called a *cache hit*. If the CPU cannot find the data within the L1 cache, it then proceeds to search for it first within L2, then L3. If it cannot find the data in any of the three, it tries to access it from memory. Each failure to find data in a cache is called a *cache miss*. Failure to locate the data in memory requires the operating system to page the data into memory from a storage device.

When the CPU writes data, it is initially written to the L1 cache. Due to ongoing activity within the CPU, at some point in time, the data will be evicted from the L1 cache into the L2 cache. The data may be further evicted from L2 and placed into L3 and eventually evicted from L3 into the memory controller's write buffers where it is then written to the memory device.

In a system that does not possess persistent memory, software persists data by writing it to a non-volatile storage device such as an SSD, HDD, SAN, NAS, or a volume in the cloud. This protects data from application or system crashes. Critical data can be manually flushed using calls such as `msync()`, `fsync()`, or `fdatasync()`, which flush uncommitted dirty pages from volatile memory to the non-volatile storage device. File systems provide `fdisk` or `chkdsk` utilities to check and attempt repairs on damaged file systems if required. File systems do not protect user data from torn blocks. Applications have a responsibility to detect and recovery from this situation. That's why databases, for example, use a variety of techniques such as transactional updates, redo/undo logging, and checksums.

 Applications memory map the persistent memory address range directly into its own memory address space. Therefore, the application must assume responsibility for checking and guaranteeing data integrity. The rest of this chapter describes your responsibilities in a persistent memory environment and how to achieve data consistency and integrity.

Power-Fail Protected Domains

A computer system may include one or more CPUs, volatile or persistent memory modules, and non-volatile storage devices such as SSDs or HDDs.

System platform hardware supports the concept of a *persistence domain*, also called *power-fail protected domains*. Depending on the platform, a persistence domain may include the persistent memory controller and write queues, memory controller write queues, and CPU caches. Once data has reached the persistence domain, it may be recoverable during a process that results from a system restart. That is, if data is located within hardware write queues or buffers protected by power failure, domain applications should assume it is persistent. For example, if a power failure occurs, the data will be flushed from the power-fail protected domain using stored energy guaranteed by the platform for this purpose. Data that has not yet made it into the protected domain will be lost.

Multiple persistence domains may exist within the same system, for example, on systems with more than one physical CPU. Systems may also provide a mechanism for partitioning the platform resources for isolation. This must be done in such a way that SNIA NVM programming model behavior is assured from each compliant volume or file system. (Chapter 3 describes the programming model as it applies to operating systems and file systems. The *Detecting Platform Capabilities* section in that chapter describes the logic that applications should perform to detect platform capabilities including power failure protected domains. Later chapters provide in-depth discussions into why, how, and when applications should flush data, if required, to guarantee the data is safe within the protected domain and persistent memory.)

Volatile memory loses its contents when the computer system's power is interrupted. Just like non-volatile storage devices, persistent memory keeps its contents even in the absence of system power. Data that has been physically saved to the persistent memory media is called *data at rest*. *Data in-flight* has the following meanings:

- Writes sent to the persistent memory device but have not yet been physically committed to the media

- Any writes that are in progress but not yet complete

- Data that has been temporarily buffered or cached in either the CPU caches or memory controller

When a system is gracefully rebooted or shut down, the system maintains power and can ensure all contents of the CPU caches and memory controllers are flushed such that any in-flight or uncommitted data is successfully written to persistent memory or non-volatile storage. When an unexpected power failure occurs, and assuming no uninterruptable power supply (UPS) is available, the system must have enough stored energy within the power supplies and capacitors dotted around it to flush data before the power is completely exhausted. Any data that is not flushed is lost and not recoverable.

Asynchronous DRAM Refresh (ADR) is a feature supported on Intel products which flushes the write-protected data buffers and places the DRAM in self-refresh. This process is critical during a power loss event or system crash to ensure the data is in a safe and consistent state on persistent memory. By default, ADR does not flush the processor caches. A platform that supports ADR only includes persistent memory and the memory controller's write pending queues within the persistence domain. This is the reason data in the CPU caches must be flushed by the application using the CLWB, CLFLUSHOPT, CLFLUSH, non-temporal stores, or WBINVD machine instructions.

Enhanced Asynchronous DRAM Refresh (eADR) requires that a non-maskable interrupt (NMI) routine be called to flush the CPU caches before the ADR event can begin. Applications running on an eADR platform do not need to perform flush operations because the hardware should flush the data automatically, but they are still required to perform an SFENCE operation to maintain write order correctness. Stores should be considered persistent only when they are globally visible, which the SFENCE guarantees.

Figure 2-2 shows both the ADR and eADR persistence domains.

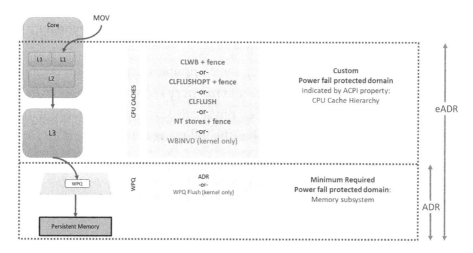

Figure 2-2.   *ADR and eADR power-fail protection domains*

ADR is a mandatory platform requirement for persistent memory. The write pending queue (WPQ) within the memory controller acknowledges receipt of the data to the writer once all the data is received. Although the data has not yet made it to the persistent media, a platform supporting ADR guarantees that it will be successfully written should a power loss event occur. During a crash or power failure, data that is in-flight through the CPU caches can only be guaranteed to be flushed to persistent media if the platform supports eADR. It will be lost on platforms that only support ADR.

The challenge with extending the persistence domain to include the CPU caches is that the CPU caches are quite large and it would take considerably more energy than the capacitors in a typical power supply can practically provide. This means the platform would have to contain batteries or utilize an external uninterruptable power supply. Requiring a battery for every server supporting persistent memory is not generally practical or cost-effective. The lifetime of a battery is typically shorter than the server, which introduces additional maintenance routines that reduce server uptime. There is also an environmental impact when using batteries as they must be disposed of or recycled correctly. It is entirely possible for server or appliance OEMs to include a battery in their product.

Because some appliance and server vendors plan to use batteries, and because platforms will someday include the CPU caches in the persistence domain, a property is available within ACPI such that the BIOS can notify software when the CPU flushes can be skipped. On platforms with eADR, there is no need for manual cache line flushing.

The Need for Flushing, Ordering, and Fencing

Except for WBINVD, which is a kernel-mode-only operation, the machine instructions in Table 2-1 (in the "Intel Machine Instructions for Persistent Memory" section) are supported in user space by Intel and AMD CPUs. Intel adopted the SNIA NVM programming model for working with persistent memory. This model allows for direct access (DAX) using byte-addressable operations (i.e., load/store). However, the persistence of the data in the cache is not guaranteed until it has entered the persistence domain. The x86 architecture provides a set of instructions for flushing cache lines in a more optimized way. In addition to existing x86 instructions, such as non-temporal stores, CLFLUSH, and WBINVD, two new instructions were added: CLFLUSHOPT and CLWB. Both new instructions must be followed by an SFENCE to ensure all flushes are completed before continuing. Flushing a cache line using CLWB, CLFLUSHOPT, or CLFLUSH and using non-temporal stores are all supported from user space. You can find details for each machine instruction in the software developer manuals for the architecture. On Intel platforms, for example, this information can be found in the Intel 64 and 32 Architectures Software Developer Manuals (https://software.intel.com/en-us/articles/intel-sdm).

Non-temporal stores imply that the data being written is not going to be read again soon, so we bypass the CPU caches. That is, there is no *temporal locality*, so there is no benefit to keeping the data in the processor's cache(s), and there may be a penalty if the stored data displaces other useful data from the cache(s).

Flushing to persistent memory directly from user space negates calling into the kernel, which makes it highly efficient. The feature is documented in the SNIA persistent memory programming model specification as an *optimized flush*. The specification document[1] describes optimized flush as optionally supported by the platform, depending on the hardware and operating system support. Despite the CPU support, it is essential for applications to use only optimized flushes when the operating system indicates that it is safe to use. The operating system may require the control point provided by calls like msync() when, for example, there are changes to file system metadata that need to be written as part of the msync() operation.

To better understand instruction ordering, consider a very simple linked list example. Our pseudocode described in the following has three simple steps to add a new node into an existing list that already contains two nodes. These steps are depicted in Figure 2-3.

1. Create the new node (Node 2).

2. Update the node pointer (next pointer) to point to the last node in the list (Node 2 → Node 1).

3. Update the head pointer to point at the new node (Head → Node 2).

Figure 2-3 (Step 3) shows that the head pointer was updated in the CPU cached version, but the Node 2 to Node 1 pointer has not yet been updated in persistent memory. This is because the hardware can choose which cache lines to commit and the order may not match the source code flow. If the system or application were to crash at this point, the persistent memory state would be inconsistent, and the data structure would no longer be usable.

---

[1]SNIA NVM programming model spec: https://www.snia.org/tech_activities/standards/ curr_standards/npm

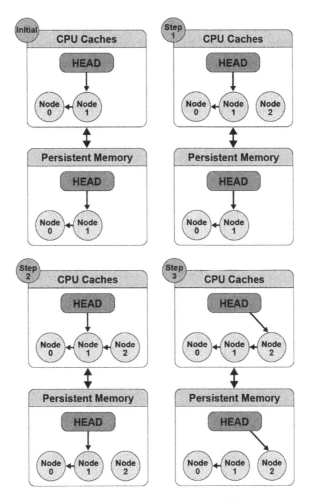

Figure 2-3. *Adding a new node to an existing linked list without a store barrier*

To solve this problem, we introduce a memory store barrier to ensure the order of the write operations is maintained. Starting from the same initial state, the pseudocode now looks like this:

1. Create the new node.

2. Update the node pointer (next pointer) to point to the last node in the list, and perform a store barrier/fence operation.

3. Update the head pointer to point at the new node.

Figure 2-4 shows that the addition of the store barrier allows the code to work as expected and maintains a consistent data structure in the volatile CPU caches and on persistent memory. We can see in Step 3 that the store barrier/fence operation waited for the pointer from Node 2 to Node 1 to update before updating the head pointer. The updates in the CPU cache matches the persistent memory version, so it now globally visible. This is a simplistic approach to solving the problem because store barriers do not provide atomicity or data integrity. A complete solution should also use transactions to ensure the data is atomically updated.

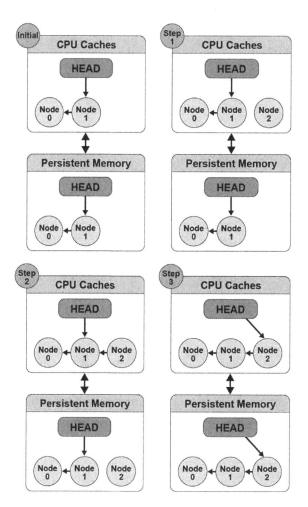

Figure 2-4. *Adding a new node to an existing linked list using a store barrier*

The PMDK detects the platform, CPU, and persistent memory features when the memory pool is opened and then uses the optimal instructions and fencing to preserve write ordering. (Memory pools are files that are memory mapped into the process address space; later chapters describe them in more detail.)

To insulate application developers from the complexities of the hardware and to keep them from having to research and implement code specific to each platform or device, the libpmem library provides a function that tells the application when optimized flush is safe to use or fall back to the standard way of flushing stores to memory-mapped files.

To simplify programming, we encourage developers to use libraries, such as libpmem and others within the PMDK. The libpmem library is also designed to detect the case of the platform with a battery that automatically converts flush calls into simple SFENCE instructions. Chapter 5 introduces and describes the core libraries within the PMDK in more detail, and later chapters take an in-depth look into each of the libraries to help you understand their APIs and features.

Data Visibility

When data is visible to other processes or threads, and when it is safe in the persistence domain, is critical to understand when using persistent memory in applications. In the Figure 2-2 and 2-3 examples, updates made to data in the CPU caches could become visible to other processes or threads. Visibility and persistence are often not the same thing, and changes made to persistent memory are often visible to other running threads in the system before they are persistent. Visibility works the same way as it does for normal DRAM, described by the memory model ordering and visibility rules for a given platform (for example, see the Intel Software Development Manual for the visibility rules for Intel platforms). Persistence of changes is achieved in one of three ways: either by calling the standard storage API for persistence (msync on Linux or FlushFileBuffers on Windows), by using optimized flush when supported, or by achieving visibility on a platform where the CPU caches are considered persistent. This is one reason we use flushing and fencing operations.

A pseudo C code example may look like this:

```
open()   // Open a file on a file system
...
mmap()   // Memory map the file
...
strcpy() // Execute a store operation
...      // Data is globally visible
msync()  // Data is now persistent
```

Developing for persistent memory follows this decades-old model.

Intel Machine Instructions for Persistent Memory

Applicable to Intel- and AMD-based ADR platforms, executing an Intel 64 and 32 architecture store instruction is not enough to make data persistent since the data may be sitting in the CPU caches indefinitely and could be lost by a power failure. Additional cache flush actions are required to make the stores persistent. Importantly, these non-privileged cache flush operations can be called from user space, meaning applications decide when and where to fence and flush data. Table 2-1 summarizes each of these instructions. For more detailed information, the Intel 64 and 32 Architectures Software Developer Manuals are online at https://software.intel.com/en-us/articles/intel-sdm.

Developers should primarily focus on CLWB and Non-Temporal Stores if available and fall back to the others as necessary. Table 2-1 lists other opcodes for completeness.

Table 2-1. *Intel architecture instructions for persistent memory*

OPCODE	Description
CLFLUSH	This instruction, supported in many generations of CPU, flushes a single cache line. Historically, this instruction is serialized, causing multiple CLFLUSH instructions to execute one after the other, without any concurrency.
CLFLUSHOPT (followed by an SFENCE)	This instruction, newly introduced for persistent memory support, is like CLFLUSH but without the serialization. To flush a range, the software executes a CLFLUSHOPT instruction for each 64-byte cache line in the range, followed by a single SFENCE instruction to ensure the flushes are complete before continuing. CLFLUSHOPT is optimized, hence the name, to allow some concurrency when executing multiple CLFLUSHOPT instructions back-to-back.
CLWB (followed by an SFENCE)	The effect of cache line writeback (CLWB) is the same as CLFLUSHOPT except that the cache line may remain valid in the cache but is no longer dirty since it was flushed. This makes it more likely to get a cache hit on this line if the data is accessed again later.
Non-temporal stores (followed by an SFENCE)	This feature has existed for a while in x86 CPUs. These stores are "write combining" and bypass the CPU cache; using them does not require a flush. A final SFENCE instruction is still required to ensure the stores have reached the persistence domain.

(continued)

Table 2-1. (*continued*)

OPCODE	Description
SFENCE	Performs a serializing operation on all store-to-memory instructions that were issued prior to the SFENCE instruction. This serializing operation guarantees that every store instruction that precedes in program order the SFENCE instruction is globally visible before any store instruction that follows the SFENCE instruction can be globally visible. The SFENCE instruction is ordered with respect to store instructions, other SFENCE instructions, any MFENCE instructions, and any serializing instructions (such as the CPUID instruction). It is not ordered with respect to load instructions or the LFENCE instruction.
WBINVD	This kernel-mode-only instruction flushes and invalidates every cache line on the CPU that executes it. After executing this on all CPUs, all stores to persistent memory are certainly in the persistence domain, but all cache lines are empty, impacting performance. Also, the overhead of sending a message to each CPU to execute this instruction can be significant. Because of this, WBINVD is only expected to be used by the kernel for flushing very large ranges (at least many megabytes).

Detecting Platform Capabilities

Server platform, CPU, and persistent memory features and capabilities are exposed to the operating system through the BIOS and ACPI that can be queried by applications. Applications should not assume they are running on hardware with all the optimizations available. Even if the physical hardware supports it, virtualization technologies may or may not expose those features to the guests, or your operating system may or may not implement them. As such, we encourage developers to use libraries, such as those in the PMDK, that perform the required feature checks or implement the checks within the application code base.

Figure 2-5 shows the flow implemented by libpmem, which initially verifies the memory-mapped file (called a memory pool), resides on a file system that has the DAX feature enabled, and is backed by physical persistent memory. Chapter 3 describes DAX in more detail.

On Linux, direct access is achieved by mounting an XFS or ext4 file system with the "-o dax" option. On Microsoft Windows, NTFS enables DAX when the volume is created and formatted using the DAX option. If the file system is not DAX-enabled, applications should fall back to the legacy approach of using msync(), fsync(), or FlushFileBuffers(). If the file system is DAX-enabled, the next check is to determine whether the platform supports ADR or eADR by verifying whether or not the CPU caches are considered persistent. On an eADR platform where CPU caches are considered persistent, no further action is required. Any data written will be considered persistent, and thus there is no requirement to perform any flushes, which is a significant performance optimization. On an ADR platform, the next sequence of events identifies the most optimal flush operation based on Intel machine instructions previously described.

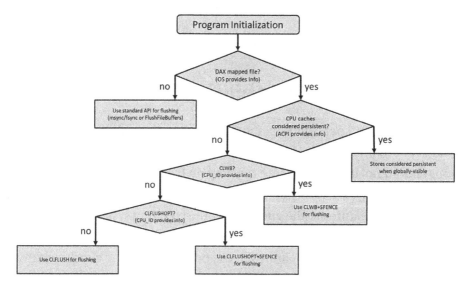

Figure 2-5. *Flowchart showing how applications can detect platform features*

Application Startup and Recovery

In addition to detecting platform features, applications should verify whether the platform was previously stopped and restarted gracefully or ungracefully. Figure 2-6 shows the checks performed by the Persistent Memory Development Kit.

Some persistent memory devices, such as Intel Optane DC persistent memory, provide SMART counters that can be queried to check the health and status. Several libraries such as `libpmemobj` query the BIOS, ACPI, OS, and persistent memory module information then perform the necessary validation steps to decide which flush operation is most optimal to use.

We described earlier that if a system loses power, there should be enough stored energy within the power supplies and platform to successfully flush the contents of the memory controller's WPQ and the write buffers on the persistent memory devices. Data will be considered consistent upon successful completion. If this process fails, due to exhausting all the stored energy before all the data was successfully flushed, the persistent memory modules will report a *dirty shutdown*. A dirty shutdown indicates that data on the device may be inconsistent. This may or may not result in needing to restore the data from backups. You can find more information on this process – and what errors and signals are sent – in the RAS (reliability, availability, serviceability) documentation for your platform and the persistent memory device. Chapter 17 also discusses this further.

Assuming no dirty shutdown is indicated, the application should check to see if the persistent memory media is reporting any known poison blocks (see Figure 2-6). Poisoned blocks are areas on the physical media that are known to be bad.

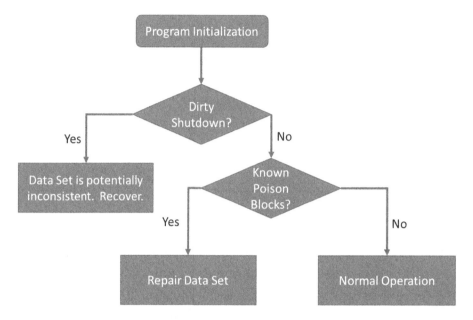

Figure 2-6. *Application startup and recovery flow*

If an application were not to check these things at startup, due to the persistent nature of the media, it could get stuck in an infinite loop, for example:

1. Application starts.

2. Reads a memory address.

3. Encounters poison.

4. Crashes or system crashes and reboots.

5. Starts and resumes operation from where it left off.

6. Performs a read on the same memory address that triggered the previous restart.

7. Application or system crashes.

8. ...

9. Repeats infinitely until manual intervention.

The ACPI specification defines an Address Range Scrub (ARS) operation that the operating system implements. This allows the operating system to perform a runtime background scan operation across the memory address range of the persistent memory.

System administrators may manually initiate an ARS. The intent is to identify bad or potentially bad memory regions before the application does. If ARS identifies an issue, the hardware can provide a status notification to the operating system and the application that can be consumed and handled gracefully. If the bad address range contains data, some method to reconstruct or restore the data needs to be implemented. Chapter 17 describes ARS in more detail.

Developers are free to implement these features directly within the application code. However, the libraries in the PMDK handle these complex conditions, and they will be maintained for each product generation while maintaining stable APIs. This gives you a future-proof option without needing to understand the intricacies of each CPU or persistent memory product.

What's Next?

Chapter 3 continues to provide foundational information from the perspective of the kernel and user spaces. We describe how operating systems such as Linux and Windows have adopted and implemented the SNIA non-volatile programming model that defines recommended behavior between various user space and operating system kernel components supporting persistent memory. Later chapters build on the foundations provided in Chapters 1 through 3.

Summary

This chapter defines persistent memory and its characteristics, recaps how CPU caches work, and describes why it is crucial for applications directly accessing persistent memory to assume responsibility for flushing CPU caches. We focus primarily on hardware implementations. User libraries, such as those delivered with the PMDK, assume the responsibilities for architecture and hardware-specific operations and allow developers to use simple APIs to implement them. Later chapters describe the PMDK libraries in more detail and show how to use them in your application.

CHAPTER 3

Operating System Support for Persistent Memory

This chapter describes how operating systems manage persistent memory as a platform resource and describes the options they provide for applications to use persistent memory. We first compare memory and storage in popular computer architectures and then describe how operating systems have been extended for persistent memory.

Operating System Support for Memory and Storage

Figure 3-1 shows a simplified view of how operating systems manage storage and volatile memory. As shown, the volatile main memory is attached directly to the CPU through a memory bus. The operating system manages the mapping of memory regions directly into the application's visible memory address space. Storage, which usually operates at speeds much slower than the CPU, is attached through an I/O controller. The operating system handles access to the storage through device driver modules loaded into the operating system's I/O subsystem.

© The Author(s) 2020

S. Scargall, *Programming Persistent Memory*, https://doi.org/10.1007/978-1-4842-4932-1_3

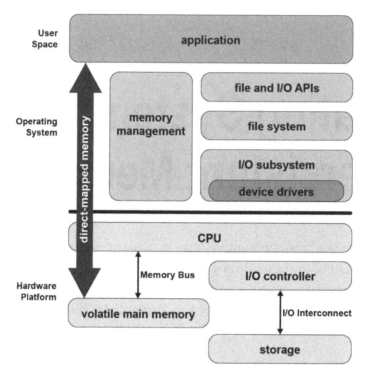

Figure 3-1. *Storage and volatile memory in the operating system*

The combination of direct application access to volatile memory combined with the operating system I/O access to storage devices supports the most common application programming model taught in introductory programming classes. In this model, developers allocate data structures and operate on them at byte granularity in memory. When the application wants to save data, it uses standard file API system calls to write the data to an open file. Within the operating system, the file system executes this write by performing one or more I/O operations to the storage device. Because these I/O operations are usually much slower than CPU speeds, the operating system typically suspends the application until the I/O completes.

Since persistent memory can be accessed directly by applications and can persist data in place, it allows operating systems to support a new programming model that combines the performance of memory while persisting data like a non-volatile storage device. Fortunately for developers, while the first generation of persistent memory was under development, Microsoft Windows and Linux designers, architects and developers collaborated in the Storage and Networking Industry Association (SNIA) to define a common programming model, so the methods for using persistent memory described in this chapter are available in both operating systems. More details can be found in the SNIA NVM programming model specification (`https://www.snia.org/tech_activities/standards/curr_standards/npm`).

Persistent Memory As Block Storage

The first operating system extension for persistent memory is the ability to detect the existence of persistent memory modules and load a device driver into the operating system's I/O subsystem as shown in Figure 3-2. This NVDIMM driver serves two important functions. First, it provides an interface for management and system administrator utilities to configure and monitor the state of the persistent memory hardware. Second, it functions similarly to the storage device drivers.

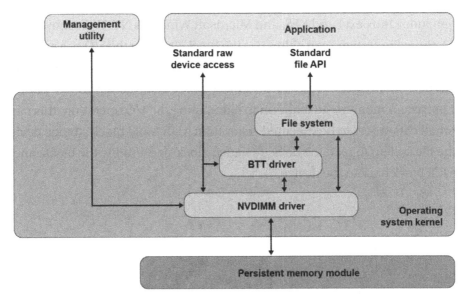

Figure 3-2. *Persistent memory as block storage*

The NVDIMM driver presents persistent memory to applications and operating system modules as a fast block storage device. This means applications, file systems, volume managers, and other storage middleware layers can use persistent memory the same way they use storage today, without modifications.

Figure 3-2 also shows the Block Translation Table (BTT) driver, which can be optionally configured into the I/O subsystem. Storage devices such as HDDs and SSDs present a native block size with 512k and 4k bytes as two common native block sizes. Some storage devices, especially NVM Express SSDs, provide a guarantee that when a power failure or server failure occurs while a block write is in-flight, either all or none of the block will be written. The BTT driver provides the same guarantee when using persistent memory as a block storage device. Most applications and file systems depend on this atomic write guarantee and should be configured to use the BTT driver, although operating systems also provide the option to bypass the BTT driver for applications that implement their own protection against partial block updates.

Persistent Memory-Aware File Systems

The next extension to the operating system is to make the file system aware of and be optimized for persistent memory. File systems that have been extended for persistent memory include Linux ext4 and XFS, and Microsoft Windows NTFS. As shown in Figure 3-3, these file systems can either use the block driver in the I/O subsystem (as described in the previous section) or bypass the I/O subsystem to directly use persistent memory as byte-addressable load/store memory as the fastest and shortest path to data stored in persistent memory. In addition to eliminating the I/O operation, this path enables small data writes to be executed faster than traditional block storage devices that require the file system to read the device's native block size, modify the block, and then write the full block back to the device.

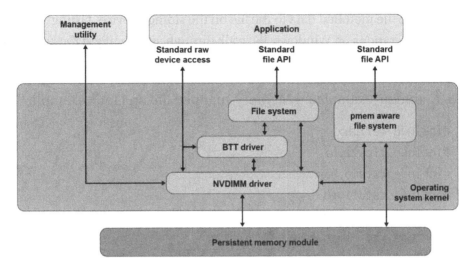

Figure 3-3. *Persistent memory-aware file system*

These persistent memory-aware file systems continue to present the familiar, standard file APIs to applications including the open, close, read, and write system calls. This allows applications to continue using the familiar file APIs while benefiting from the higher performance of persistent memory.

Memory-Mapped Files

Before describing the next operating system option for using persistent memory, this section reviews memory-mapped files in Linux and Windows. When memory mapping a file, the operating system adds a range to the application's virtual address space which corresponds to a range of the file, paging file data into physical memory as required. This allows an application to access and modify file data as byte-addressable in-memory data structures. This has the potential to improve performance and simplify application development, especially for applications that make frequent, small updates to file data.

Applications memory map a file by first opening the file, then passing the resulting file handle as a parameter to the mmap() system call in Linux or to MapViewOfFile() in Windows. Both return a pointer to the in-memory copy of a portion of the file. Listing 3-1 shows an example of Linux C code that memory maps a file, writes data into the file by accessing it like memory, and then uses the msync system call to perform the I/O operation to write the modified data to the file on the storage device. Listing 3-2 shows the equivalent operations on Windows. We walk through and highlight the key steps in both code samples.

Listing 3-1. mmap_example.c – Memory-mapped file on Linux example

```
50   #include <err.h>
51   #include <fcntl.h>
52   #include <stdio.h>
53   #include <stdlib.h>
54   #include <string.h>
55   #include <sys/mman.h>
56   #include <sys/stat.h>
57   #include <sys/types.h>
58   #include <unistd.h>
59
60   int
61   main(int argc, char *argv[])
62   {
63       int fd;
64       struct stat stbuf;
65       char *pmaddr;
66
67       if (argc != 2) {
68           fprintf(stderr, "Usage: %s filename\n",
69               argv[0]);
70           exit(1);
71       }
72
73       if ((fd = open(argv[1], O_RDWR)) < 0)
74           err(1, "open %s", argv[1]);
75
76       if (fstat(fd, &stbuf) < 0)
77           err(1, "stat %s", argv[1]);
78
79       /*
```

```
80          * Map the file into our address space for read
81          * & write. Use MAP_SHARED so stores are visible
82          * to other programs.
83          */
84         if ((pmaddr = mmap(NULL, stbuf.st_size,
85                     PROT_READ|PROT_WRITE,
86                     MAP_SHARED, fd, 0)) == MAP_FAILED)
87             err(1, "mmap %s", argv[1]);
88
89         /* Don't need the fd anymore because the mapping
90          * stays around */
91         close(fd);
92
93         /* store a string to the Persistent Memory */
94         strcpy(pmaddr, "This is new data written to the
95                 file");
96
97         /*
98          * Simplest way to flush is to call msync().
99          * The length needs to be rounded up to a 4k page.
100         */
101        if (msync((void *)pmaddr, 4096, MS_SYNC) < 0)
102            err(1, "msync");
103
104        printf("Done.\n");
105        exit(0);
106    }
```

- Lines 67-74: We verify the caller passed a file name that can be opened. The open call will create the file if it does not already exist.

- Line 76: We retrieve the file statistics to use the length when we memory map the file.

- Line 84: We map the file into the application's address space to allow our program to access the contents as if in memory. In the second parameter, we pass the length of the file, requesting Linux to initialize memory with the full file. We also map the file with both READ and WRITE access and also as SHARED allowing other processes to map the same file.

- Line 91: We retire the file descriptor which is no longer needed once a file is mapped.

- Line 94: We write data into the file by accessing it like memory through the pointer returned by mmap.

- Line 101: We explicitly flush the newly written string to the backing storage device.

Listing 3-2 shows an example of C code that memory maps a file, writes data into the file, and then uses the FlushViewOfFile() and FlushFileBuffers() system calls to flush the modified data to the file on the storage device.

Listing 3-2. Memory-mapped file on Windows example

```
45  #include <fcntl.h>
46  #include <stdio.h>
47  #include <stdlib.h>
48  #include <string.h>
49  #include <sys/stat.h>
50  #include <sys/types.h>
51  #include <Windows.h>
52
53  int
54  main(int argc, char *argv[])
55  {
56      if (argc != 2) {
57          fprintf(stderr, "Usage: %s filename\n",
58              argv[0]);
59          exit(1);
60      }
61
```

```
62      /* Create the file or open if the file exists */
63      HANDLE fh = CreateFile(argv[1],
64          GENERIC_READ|GENERIC_WRITE,
65          0,
66          NULL,
67          OPEN_EXISTING,
68          FILE_ATTRIBUTE_NORMAL,
69          NULL);
70
71      if (fh == INVALID_HANDLE_VALUE) {
72          fprintf(stderr, "CreateFile, gle: 0x%08x",
73              GetLastError());
74          exit(1);
75      }
76
77      /*
78       * Get the file length for use when
79       * memory mapping later
80       * */
81      DWORD filelen = GetFileSize(fh, NULL);
82      if (filelen == 0) {
83          fprintf(stderr, "GetFileSize, gle: 0x%08x",
84              GetLastError());
85          exit(1);
86      }
87
88      /* Create a file mapping object */
89      HANDLE fmh = CreateFileMapping(fh,
90          NULL, /* security attributes */
91          PAGE_READWRITE,
92          0,
93          0,
94          NULL);
95
```

```
 96        if (fmh == NULL) {
 97            fprintf(stderr, "CreateFileMapping,
 98                gle: 0x%08x", GetLastError());
 99            exit(1);
100        }
101
102        /*
103         * Map into our address space and get a pointer
104         * to the beginning
105         * */
106        char *pmaddr = (char *)MapViewOfFileEx(fmh,
107            FILE_MAP_ALL_ACCESS,
108            0,
109            0,
110            filelen,
111            NULL); /* hint address */
112
113        if (pmaddr == NULL) {
114            fprintf(stderr, "MapViewOfFileEx,
115                gle: 0x%08x", GetLastError());
116            exit(1);
117        }
118
119        /*
120         * On windows must leave the file handle(s)
121         * open while mmaped
122         * */
123
124        /* Store a string to the beginning of the file  */
125        strcpy(pmaddr, "This is new data written to
126            the file");
127
128        /*
129         * Flush this page with length rounded up to 4K
130         * page size
131         * */
```

```
132     if (FlushViewOfFile(pmaddr, 4096) == FALSE) {
133         fprintf(stderr, "FlushViewOfFile,
134             gle: 0x%08x", GetLastError());
135         exit(1);
136     }
137
138     /* Flush the complete file to backing storage */
139     if (FlushFileBuffers(fh) == FALSE) {
140         fprintf(stderr, "FlushFileBuffers,
141             gle: 0x%08x", GetLastError());
142         exit(1);
143     }
144
145     /* Explicitly unmap before closing the file */
146     if (UnmapViewOfFile(pmaddr) == FALSE) {
147         fprintf(stderr, "UnmapViewOfFile,
148             gle: 0x%08x", GetLastError());
149         exit(1);
150     }
151
152     CloseHandle(fmh);
153     CloseHandle(fh);
154
155     printf("Done.\n");
156     exit(0);
157 }
```

- Lines 45-75: As in the previous Linux example, we take the file name passed through argv and open the file.

- Line 81: We retrieve the file size to use later when memory mapping.

- Line 89: We take the first step to memory mapping a file by creating the file mapping. This step does not yet map the file into our application's memory space.

- Line 106: This step maps the file into our memory space.

- Line 125: As in the previous Linux example, we write a string to the beginning of the file, accessing the file like memory.

- Line 132: We flush the modified memory page to the backing storage.

- Line 139: We flush the full file to backing storage, including any additional file metadata maintained by Windows.

- Line 146-157: We unmap the file, close the file, then exit the program.

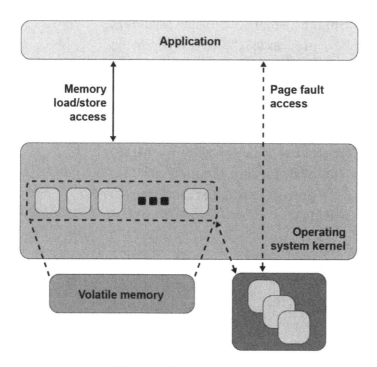

Figure 3-4. *Memory-mapped files with storage*

Figure 3-4 shows what happens inside the operating system when an application calls mmap() on Linux or CreateFileMapping() on Windows. The operating system allocates memory from its memory page cache, maps that memory into the application's address space, and creates the association with the file through a storage device driver.

As the application reads pages of the file in memory, and if those pages are not present in memory, a page fault exception is raised to the operating system which will then read that page into main memory through storage I/O operations. The operating system also tracks writes to those memory pages and schedules asynchronous I/O operations to write the modifications back to the primary copy of the file on the storage device. Alternatively, if the application wants to ensure updates are written back to storage before continuing as we did in our code example, the msync system call on Linux or FlushViewOfFile on Windows executes the flush to disk. This may cause the operating system to suspend the program until the write finishes, similar to the file-write operation described earlier.

This description of memory-mapped files using storage highlights some of the disadvantages. First, a portion of the limited kernel memory page cache in main memory is used to store a copy of the file. Second, for files that cannot fit in memory, the application may experience unpredictable and variable pauses as the operating system moves pages between memory and storage through I/O operations. Third, updates to the in-memory copy are not persistent until written back to storage so can be lost in the event of a failure.

Persistent Memory Direct Access (DAX)

The persistent memory direct access feature in operating systems, referred to as DAX in Linux and Windows, uses the memory-mapped file interfaces described in the previous section but takes advantage of persistent memory's native ability to both store data and to be used as memory. Persistent memory can be natively mapped as application memory, eliminating the need for the operating system to cache files in volatile main memory.

To use DAX, the system administrator creates a file system on the persistent memory module and mounts that file system into the operating system's file system tree. For Linux users, persistent memory devices will appear as /dev/pmem* device special files. To show the persistent memory physical devices, system administrators can use the ndctl and ipmctl utilities shown in Listings 3-3 and 3-4.

Listing 3-3. Displaying persistent memory physical devices and regions on Linux

```
# ipmctl show -dimm

DimmID | Capacity  | HealthState | ActionRequired | LockState | FWVersion
================================================================================
0x0001 | 252.4 GiB | Healthy     | 0              | Disabled  | 01.02.00.5367
0x0011 | 252.4 GiB | Healthy     | 0              | Disabled  | 01.02.00.5367
0x0021 | 252.4 GiB | Healthy     | 0              | Disabled  | 01.02.00.5367
0x0101 | 252.4 GiB | Healthy     | 0              | Disabled  | 01.02.00.5367
0x0111 | 252.4 GiB | Healthy     | 0              | Disabled  | 01.02.00.5367
0x0121 | 252.4 GiB | Healthy     | 0              | Disabled  | 01.02.00.5367
0x1001 | 252.4 GiB | Healthy     | 0              | Disabled  | 01.02.00.5367
0x1011 | 252.4 GiB | Healthy     | 0              | Disabled  | 01.02.00.5367
0x1021 | 252.4 GiB | Healthy     | 0              | Disabled  | 01.02.00.5367
0x1101 | 252.4 GiB | Healthy     | 0              | Disabled  | 01.02.00.5367
0x1111 | 252.4 GiB | Healthy     | 0              | Disabled  | 01.02.00.5367
0x1121 | 252.4 GiB | Healthy     | 0              | Disabled  | 01.02.00.5367

# ipmctl show -region

SocketID| ISetID          | PersistentMemoryType | Capacity   | FreeCapacity | HealthState
==========================================================================================
0x0000  | 0x2d3c7f48f4e22ccc | AppDirect         | 1512.0 GiB | 0.0 GiB      | Healthy
0x0001  | 0xdd387f488ce42ccc | AppDirect         | 1512.0 GiB | 1512.0 GiB   | Healthy
```

Listing 3-4. Displaying persistent memory physical devices, regions, and namespaces on Linux

```
# ndctl list -DRN
{
  "dimms":[
    {
      "dev":"nmem1",
      "id":"8089-a2-1837-00000bb3",
      "handle":17,
```

```
      "phys_id":44,
      "security":"disabled"
    },
    {
      "dev":"nmem3",
      "id":"8089-a2-1837-00000b5e",
      "handle":257,
      "phys_id":54,
      "security":"disabled"
    },
    [...snip...]
    {
      "dev":"nmem8",
      "id":"8089-a2-1837-00001114",
      "handle":4129,
      "phys_id":76,
      "security":"disabled"
    }
  ],
  "regions":[
    {
      "dev":"region1",
      "size":1623497637888,
      "available_size":1623497637888,
      "max_available_extent":1623497637888,
      "type":"pmem",
      "iset_id":-2506113243053544244,
      "mappings":[
        {
          "dimm":"nmem11",
          "offset":268435456,
          "length":270582939648,
          "position":5
        },
```

```
      {
        "dimm":"nmem10",
        "offset":268435456,
        "length":270582939648,
        "position":1
      },
      {
        "dimm":"nmem9",
        "offset":268435456,
        "length":270582939648,
        "position":3
      },
      {
        "dimm":"nmem8",
        "offset":268435456,
        "length":270582939648,
        "position":2
      },
      {
        "dimm":"nmem7",
        "offset":268435456,
        "length":270582939648,
        "position":4
      },
      {
        "dimm":"nmem6",
        "offset":268435456,
        "length":270582939648,
        "position":0
      }
    ],
    "persistence_domain":"memory_controller"
  },
  {
    "dev":"region0",
    "size":1623497637888,
```

```
"available_size":0,
"max_available_extent":0,
"type":"pmem",
"iset_id":3259620181632232652,
"mappings":[
  {
    "dimm":"nmem5",
    "offset":268435456,
    "length":270582939648,
    "position":5
  },
  {
    "dimm":"nmem4",
    "offset":268435456,
    "length":270582939648,
    "position":1
  },
  {
    "dimm":"nmem3",
    "offset":268435456,
    "length":270582939648,
    "position":3
  },
  {
    "dimm":"nmem2",
    "offset":268435456,
    "length":270582939648,
    "position":2
  },
  {
    "dimm":"nmem1",
    "offset":268435456,
    "length":270582939648,
    "position":4
  },
```

```
      {
        "dimm":"nmem0",
        "offset":268435456,
        "length":270582939648,
        "position":0
      }
    ],
    "persistence_domain":"memory_controller",
    "namespaces":[
      {
        "dev":"namespace0.0",
        "mode":"fsdax",
        "map":"dev",
        "size":1598128390144,
        "uuid":"06b8536d-4713-487d-891d-795956d94cc9",
        "sector_size":512,
        "align":2097152,
        "blockdev":"pmem0"
      }
    ]
  }
 ]
}
```

When a file system is created and mounted using /dev/pmem* devices, they can be identified using the df command as shown in Listing 3-5.

Listing 3-5. Locating persistent memory on Linux.

```
$ df -h /dev/pmem*
Filesystem      Size  Used Avail Use% Mounted on
/dev/pmem0      1.5T   77M  1.4T   1% /mnt/pmemfs0
/dev/pmem1      1.5T   77M  1.4T   1% /mnt/pmemfs1
```

Windows developers will use PowerShellCmdlets as shown in Listing 3-6. In either case, assuming the administrator has granted you rights to create files, you can create one or more files in the persistent memory and then memory map those files to your application using the same method shown in code Listings 3-1 and 3-2.

Listing 3-6. Locating persistent memory on Windows

```
PS C:\Users\Administrator> Get-PmemDisk

Number Size    Health  Atomicity Removable Physical device IDs Unsafe shutdowns
------ ----    ------  --------- --------- ------------------- ----------------
2      249 GB Healthy None      True      {1}                 36

PS C:\Users\Administrator> Get-Disk 2 | Get-Partition

PartitionNumber DriveLetter Offset   Size      Type
--------------- ----------- ------   ----      ----
1                           24576    15.98 MB  Reserved
2               D           16777216 248.98 GB Basic
```

Managing persistent memory as files has several benefits:

- You can leverage the rich features of leading file systems for organizing, managing, naming, and limiting access for user's persistent memory files and directories.

- You can apply the familiar file system permissions and access rights management for protecting data stored in persistent memory and for sharing persistent memory between multiple users.

- System administrators can use existing backup tools that rely on file system revision-history tracking.

- You can build on existing memory mapping APIs as described earlier and applications that currently use memory-mapped files and can use direct persistent memory without modifications.

Once a file backed by persistent memory is created and opened, an application still calls mmap() or MapViewOfFile() to get a pointer to the persistent media. The difference, shown in Figure 3-5, is that the persistent memory-aware file system recognizes that the file is on persistent memory and programs the memory management unit (MMU) in the CPU to map the persistent memory directly into the application's address space. Neither a copy in kernel memory nor synchronizing to storage through I/O operations is required. The application can use the pointer returned by mmap() or MapViewOfFile() to operate on its data in place directly in the persistent memory. Since no kernel I/O operations are required, and because the full file is mapped into the application's memory, it can manipulate large collections of data objects with higher and more consistent performance as compared to files on I/O-accessed storage.

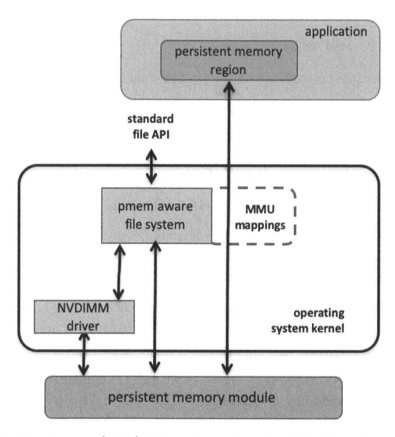

Figure 3-5. *Direct access (DAX) I/O and standard file API I/O paths through the kernel*

Listing 3-7 shows a C source code example that uses DAX to write a string directly into persistent memory. This example uses one of the persistent memory API libraries included in Linux and Windows called libpmem. Although we discuss these libraries in depth in later chapters, we describe the use of two of the functions available in libpmem in the following steps. The APIs in libpmem are common across Linux and Windows and abstract the differences between underlying operating system APIs, so this sample code is portable across both operating system platforms.

Listing 3-7. DAX programming example

```
32   #include <sys/types.h>
33   #include <sys/stat.h>
34   #include <fcntl.h>
35   #include <stdio.h>
36   #include <errno.h>
37   #include <stdlib.h>
38   #ifndef _WIN32
39   #include <unistd.h>
40   #else
41   #include <io.h>
42   #endif
43   #include <string.h>
44   #include <libpmem.h>
45
46   /* Using 4K of pmem for this example */
47   #define PMEM_LEN 4096
48
49   int
50   main(int argc, char *argv[])
51   {
52       char *pmemaddr;
53       size_t mapped_len;
54       int is_pmem;
55
56       if (argc != 2) {
57           fprintf(stderr, "Usage: %s filename\n",
58               argv[0]);
59           exit(1);
60       }
61
62       /* Create a pmem file and memory map it. */
63       if ((pmemaddr = pmem_map_file(argv[1], PMEM_LEN,
64               PMEM_FILE_CREATE, 0666, &mapped_len,
65               &is_pmem)) == NULL) {
```

```
66              perror("pmem_map_file");
67              exit(1);
68          }
69
70          /* Store a string to the persistent memory. */
71          char s[] = "This is new data written to the file";
72          strcpy(pmemaddr, s);
73
74          /* Flush our string to persistence. */
75          if (is_pmem)
76              pmem_persist(pmemaddr, sizeof(s));
77          else
78              pmem_msync(pmemaddr, sizeof(s));
79
80          /* Delete the mappings. */
81          pmem_unmap(pmemaddr, mapped_len);
82
83          printf("Done.\n");
84          exit(0);
85      }
```

- Lines 38-42: We handle the differences between Linux and Windows for the include files.

- Line 44: We include the header file for the libpmem API used in this example.

- Lines 56-60: We take the pathname argument from the command line argument.

- Line 63-68: The pmem_map_file function in libpmem handles opening the file and mapping it into our address space on both Windows and Linux. Since the file resides on persistent memory, the operating system programs the hardware MMU in the CPU to map the persistent memory region into our application's virtual address space. Pointer pmemaddr is set to the beginning of that region. The pmem_map_file function can also be used for memory mapping disk-based files through kernel main memory as well as directly mapping persistent memory, so is_pmem is set to TRUE if the file resides on persistent memory and FALSE if mapped through main memory.

- Line 72: We write a string into persistent memory.

- Lines 75-78: If the file resides on persistent memory, the pmem_persist function uses the user space machine instructions (described in Chapter 2) to ensure our string is flushed through CPU cache levels to the power-fail safe domain and ultimately to persistent memory. If our file resided on disk-based storage, Linux mmap or Windows FlushViewOfFile would be used to flushed to storage. Note that we can pass small sizes here (the size of the string written is used in this example) instead of requiring flushes at page granularity when using msync() or FlushViewOfFile().

- Line 81: Finally, we unmap the persistent memory region.

Summary

Figure 3-6 shows the complete view of the operating system support that this chapter describes. As we discussed, an application can use persistent memory as a fast SSD, more directly through a persistent memory-aware file system, or mapped directly into the application's memory space with the DAX option. DAX leverages operating system services for memory-mapped files but takes advantage of the server hardware's ability to map persistent memory directly into the application's address space. This avoids the need to move data between main memory and storage. The next few chapters describe considerations for working with data directly in persistent memory and then discuss the APIs for simplifying development.

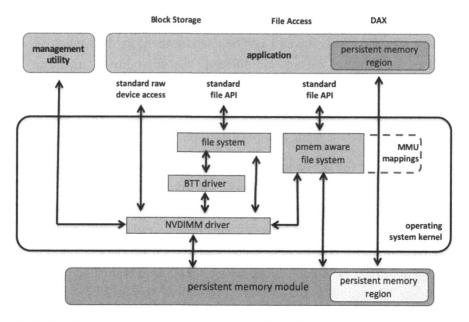

Figure 3-6. *Persistent memory programming interfaces*

## CHAPTER 4

Fundamental Concepts of Persistent Memory Programming

In Chapter 3, you saw how operating systems expose persistent memory to applications as memory-mapped files. This chapter builds on this fundamental model and examines the programming challenges that arise. Understanding these challenges is an essential part of persistent memory programming, especially when designing a strategy for recovery after application interruption due to issues like crashes and power failures. However, do not let these challenges deter you from persistent memory programming! Chapter 5 describes how to leverage existing solutions to save you programming time and reduce complexity.

What's Different?

Application developers typically think in terms of *memory-resident* data structures and *storage-resident* data structures. For data center applications, developers are careful to maintain consistent data structures on storage, even in the face of a system crash. This problem is commonly solved using logging techniques such as *write-ahead logging*, where changes are first written to a log and then flushed to persistent storage. If the data modification process is interrupted, the application has enough information in the log to finish the operation on restart. Techniques like this have been around for many years; however, correct implementations are challenging to develop and time-consuming to maintain. Developers often rely on a combination of databases, libraries, and modern file systems to provide consistency. Even so, it is ultimately the application developer's

© The Author(s) 2020
S. Scargall, *Programming Persistent Memory*, https://doi.org/10.1007/978-1-4842-4932-1_4

responsibility to design in a strategy to maintain consistent data structures on storage, both at runtime and when recovering from application and system crashes.

Unlike storage-resident data structures, application developers are concerned about maintaining consistency of memory-resident data structures at runtime. When an application has multiple threads accessing the same data structure, techniques like *locking* are used so that one thread can perform complex changes to a data structure without another thread seeing only part of the change. When an application exits or crashes, or the system crashes, the memory contents are gone, so there is no need to maintain consistency of memory-resident data structures between runs of an application like there is with storage-resident data structures.

These explanations may seem obvious, but these assumptions that the storage state stays around between runs and memory contents are volatile are so fundamental in the way applications are developed that most developers don't give it much thought. What's different about persistent memory is, of course, that it is persistent, so all the considerations of both storage and memory apply. The application is responsible for maintaining consistent data structures between runs and reboots, as well as the thread-safe locking used with memory-resident data structures.

If persistent memory has these attributes and requirements just like storage, why not use code developed over the years for storage? This approach does work; using the storage APIs on persistent memory is part of the programming model we described in Chapter 3. If the existing storage APIs on persistent memory are fast enough and meet the application's needs, then no further work is necessary. But to fully leverage the advantages of persistent memory, where data structures are read and written in place on persistence and accesses happen at the byte granularity, instead of using the block storage stack, applications will want to memory map it and access it directly. This eliminates the buffer-based storage APIs in the data path.

Atomic Updates

Each platform supporting persistent memory will have a set of *native* memory operations that are atomic. On Intel hardware, the atomic persistent store is 8 bytes. Thus, if the program or system crashes while an aligned 8-byte store to persistent memory is in-flight, on recovery those 8 bytes will either contain the old contents or the new contents. The Intel processor has instructions that store more than 8 bytes, but those are not failure atomic, so they can be *torn* by events like a power failure.

Sometimes an update to a memory-resident data structure will require multiple instructions, so naturally those changes can be torn by power failure as well since power could be lost between any two instructions. Runtime locking prevents other threads from seeing a partially done change, but locking doesn't provide any failure atomicity. When an application needs to make a change that is larger than 8 bytes to persistent memory, it must construct the atomic operation by building on top of the basic atomics provided by hardware, such as the 8-byte failure atomicity provided by Intel hardware.

Transactions

Combining multiple operations into a single atomic operation is usually referred to as a *transaction*. In the database world, the acronym ACID describes the properties of a transaction: *atomicity, consistency, isolation, and durability*.

Atomicity

As described earlier, atomicity is when multiple operations are composed into a single atomic action that either happens entirely or does not happen at all, even in the face of system failure. For persistent memory, the most common techniques used are

- Redo logging, where the full change is first written to a log, so during recovery, it can be *rolled forward* if interrupted.

- Undo logging, where information is logged that allows a partially done change to be *rolled back* during recovery.

- Atomic pointer updates, where a change is made active by updating a single pointer atomically, usually changing it from pointing to old data to new data.

The preceding list is not exhaustive, and it ignores the details that can get relatively complex. One common consideration is that transactions often include memory allocation/deallocation. For example, a transaction that adds a node to a tree data structure usually includes the allocation of the new node. If the transaction is rolled back, the memory must be freed to prevent a memory leak. Now imagine a transaction that performs multiple persistent memory allocations and free operations, all of which must be part of the same atomic operation. The implementation of this transaction is clearly more complex than just writing the new value to a log or updating a single pointer.

Consistency

Consistency means that a transaction can only move a data structure from one valid state to another. For persistent memory, programmers usually find that the locking they use to make updates thread-safe often indicates consistency points as well. If it is not valid for a thread to see an intermediate state, locking prevents it from happening, and when it is safe to drop the lock, that is because it is safe for another thread to observe the current state of the data structure.

Isolation

Multithreaded (concurrent) execution is commonplace in modern applications. When making transactional updates, the isolation is what allows the concurrent updates to have the same effect as if they were executed sequentially. At runtime, isolation for persistent memory updates is typically achieved by locking. Since the memory is persistent, the isolation must be considered for transactions that were in-flight when the application was interrupted. Persistent memory programmers typically detect this situation on restart and roll partially done transactions forward or backward appropriately before allowing general-purpose threads access to the data structures.

Durability

A transaction is considered durable if it is on persistent media when it is complete. Even if the system loses power or crashes at that point, the transaction remains completed. As described in Chapter 2, this usually means the changes must be flushed from the CPU caches. This can be done using standard APIs, such as the Linux msync() call, or platform-specific instructions such as Intel's CLWB. When implementing transactions on persistent memory, pay careful attention to ensure that log entries are flushed to persistence before changes are started and flush changes to persistence before a transaction is considered complete.

Another aspect of the durable property is the ability to find the persistent information again when an application starts up. This is so fundamental to how storage works that we take it for granted. Metadata such as file names and directory names are used to find the durable state of an application on storage. For persistent memory, the same is true due to the programming model described in Chapter 3, where persistent memory is accessed by first opening a file on a direct access (DAX) file system and then memory mapping that file. However, a memory-mapped file is just a range of raw data;

how does the application find the data structures resident in that range? For persistent memory, there must be at least one well-known location of a data structure to use as a starting point. This is often referred to as a *root object* (described in Chapter 7). The root object is used by many of the higher-level libraries within PMDK to access the data.

Flushing Is Not Transactional

It is important to separate the ideas of flushing to persistence from transactional updates. Flushing changes to storage using calls like msync() or fsync() on Linux and FlushFileBuffers() on Windows have never provided transactional updates. Applications assume the responsibility for maintaining consistent storage data structures in addition to flushing changes to storage. With persistent memory, the same is true. In Chapter 3, a simple program stored a string to persistent memory and then flushed it to make sure the change was persistent. But that code was not transactional, and in the face of failure, the change could be in just about any state – from completely lost to partially lost to fully completed.

A fundamental property of caches is that they hold data temporarily for performance, but they do not typically hold data until a transaction is ready to commit. Normal system activity can cause cache pressure and evict data at any time and in any order. If the examples in Chapter 3 were interrupted by power failure, it is possible for any part of the string being stored to be lost and any part to be persistent, in any order. It is important to think of the cache flush operation as *flush anything that hasn't already been flushed* and not as *flush all my changes now*.

Finally, we showed a decision tree in Chapter 2 (Figure 2-5) where an application can determine at startup that no cache flushing is required for persistent memory. This can be the case on platforms where the CPU cache is flushed automatically on power failure, for example. Even on platforms where flush instructions are not needed, transactions are still required to keep data structures consistent in the face of failure.

Start-Time Responsibilities

In Chapter 2 (Figures 2-5 and 2-6), we showed flowcharts outlining the application's responsibilities when using persistent memory. These responsibilities included detecting platform details, available instructions, media failures, and so on. For storage, these types of things happen in the storage stack in the operating system. Persistent memory, however, allows direct access, which removes the kernel from the data path once the file is memory mapped.

As a programmer, you may be tempted to map persistent memory and start using it, as shown in the Chapter 3 examples. For production-quality programming, you want to ensure these start-time responsibilities are met. For example, if you skip the checks in Figure 2-5, you will end up with an application that flushes CPU caches even when it is not required, and that will perform poorly on hardware that does not need the flushing. If you skip the checks in Figure 2-6, you will have an application that ignores media errors and may use corrupted data resulting in unpredictable and undefined behavior.

Tuning for Hardware Configurations

When storing a large data structure to persistent memory, there are several ways to copy the data and make it persistent. You can either copy the data using the common store operations and then flush the caches (if required) or use special instructions like Intel's non-temporal store instructions that bypass the CPU caches. Another consideration is that persistent memory write performance may be slower than writing to normal memory, so you may want to take steps to store to persistent memory as efficiently as possible, by combining multiple small writes into larger changes before storing them to persistent memory. The optimal write size for persistent memory will depend on both the platform it is plugged into and the persistent memory product itself. These examples show that different platforms will have different characteristics when using persistent memory, and any production-quality application will be tuned to perform best on the intended target platforms. Naturally, one way to help with this tuning work is to leverage libraries or middleware that has already been tuned and validated.

Summary

This chapter provides an overview of the fundamental concepts of persistent memory programming. When developing an application that uses persistent memory, you must carefully consider several areas:

- Atomic updates.

- Flushing is not transactional.

- Start-time responsibilities.

- Tuning for hardware configurations.

Handling these challenges in a production-quality application requires some complex programming and extensive testing and performance analysis. The next chapter introduces the Persistent Memory Development Kit, designed to assist application developers in solving these challenges.

CHAPTER 5

Introducing the Persistent Memory Development Kit

Previous chapters introduced the unique properties of persistent memory that make it special, and you are correct in thinking that writing software for such a novel technology is complicated. Anyone who has researched or developed code for persistent memory can testify to this. To make your job easier, Intel created the Persistent Memory Development Kit (PMDK). The team of PMDK developers envisioned it to be the standard library for all things persistent memory that would provide solutions to the common challenges of persistent memory programming.

Background

The PMDK has evolved to become a large collection of open source libraries and tools for application developers and system administrators to simplify managing and accessing persistent memory devices. It was developed alongside evolving support for persistent memory in operating systems, which ensures the libraries take advantage of all the features exposed through the operating system interfaces.

The PMDK libraries build on the SNIA NVM programming model (described in Chapter 3). They extend it to varying degrees, some by simply wrapping around the primitives exposed by the operating system with easy-to-use functions and others by providing complex data structures and algorithms for use with persistent memory. This means you are responsible for making an informed decision about which level of abstraction is the best for your use case.

© The Author(s) 2020
S. Scargall, *Programming Persistent Memory*, https://doi.org/10.1007/978-1-4842-4932-1_5

Although the PMDK was created by Intel to support its hardware products, Intel is committed to ensuring the libraries and tools are both vendor and platform neutral. This means that the PMDK is not tied to Intel processors or Intel persistent memory devices. It can be made to work on any other platform that exposes the necessary interfaces through the operating system, including Linux and Microsoft Windows. We welcome and encourage contributions to PMDK from individuals, hardware vendors, and ISVs. The PMDK has a BSD 3-Clause License, allowing developers to embed it in any software, whether it's open source or proprietary. This allows you to pick and choose individual components of PMDK by integrating only the bits of code required.

The PMDK is available at no cost on GitHub (`https://github.com/pmem/pmdk`) and has a dedicated web site at `https://pmem.io`. Man pages are delivered with PMDK and are available online under each library's own page. Appendix B of this book describes how to install it on your system.

An active persistent memory community is available through Google Forums at `https://groups.google.com/forum/#!forum/pmem`. This forum allows developers, system administrators, and others with an interest in persistent memory to ask questions and get assistance. This is a great resource.

Choosing the Right Semantics

With so many libraries available within the PMDK, it is important to carefully consider your options. The PMDK offers two library categories:

1. *Volatile* libraries are for use cases that only wish to exploit the capacity of persistent memory.

2. *Persistent* libraries are for use in software that wishes to implement fail-safe persistent memory algorithms.

While you are deciding how to best solve a problem, carefully consider which category it fits into. The challenges that fail-safe persistent programs present are significantly different from volatile ones. Choosing the right approach upfront will minimize the risk of having to rewrite any code.

You may decide to use libraries from both categories for different parts of the application, depending on feature and functional requirements.

Volatile Libraries

Volatile libraries are typically simpler to use because they can fall back to dynamic random-access memory (DRAM) when persistent memory is not available. This provides a more straightforward implementation. Depending on the workload, they may also have lower overall overhead compared to similar persistent libraries because they do not need to ensure consistency of data in the presence of failures.

This section explores the available libraries for volatile use cases in applications, including what the library is and when to use it. The libraries may have overlapping situation use cases.

libmemkind

What is it?

The memkind library, called `libmemkind`, is a user-extensible heap manager built on top of `jemalloc`. It enables control of memory characteristics and partitioning of the heap between different kinds of memory. The kinds of memory are defined by operating system memory policies that have been applied to virtual address ranges. Memory characteristics supported by memkind without user extension include control of nonuniform memory access (NUMA) and page size features. The `jemalloc` nonstandard interface has been extended to enable specialized kinds to make requests for virtual memory from the operating system through the memkind partition interface. Through the other memkind interfaces, you can control and extend memory partition features and allocate memory while selecting enabled features. The memkind interface allows you to create and control file-backed memory from persistent memory with PMEM kind.

Chapter 10 describes this library in more detail. You can download memkind and read the architecture specification and API documentation at `http://memkind.github. io/memkind/`. memkind is an open source project on GitHub at `https://github.com/ memkind/memkind`.

When to use it?

Choose `libmemkind` when you want to manually move select memory objects to persistent memory in a volatile application while retaining the traditional programming model. The memkind library provides familiar `malloc()` and `free()` semantics. This is the recommended memory allocator for most volatile use cases of persistent memory.

Modern memory allocators usually rely on anonymous memory mapping to provision memory pages from the operating system. For most systems, this means that actual physical memory is allocated only when a page is first accessed, allowing the OS to overprovision virtual memory. Additionally, anonymous memory can be paged out if needed. When using memkind with file-based kinds, such as PMEM kind, physical space is still only allocated on first access to a page and the other described techniques no longer apply. Memory allocation will fail when there is no memory available to be allocated, so it is important to handle such failures within the application.

The described techniques also play an important role in hiding the inherent inefficiencies of manual dynamic memory allocation such as fragmentation, which causes allocation failures when not enough contiguous free space is available. Thus, file-based kinds can exhibit low space utilization for applications with irregular allocation/deallocation patterns. Such workloads may be better served with `libvmemcache`.

libvmemcache

What is it?

`libvmemcache` is an embeddable and lightweight in-memory caching solution that takes full advantage of large-capacity memory, such as persistent memory with direct memory access (DAX), through memory mapping in an efficient and scalable way. `libvmemcache` has unique characteristics:

- An extent-based memory allocator sidesteps the fragmentation problem that affects most in-memory databases and allows the cache to achieve very high space utilization for most workloads.

- The buffered least recently used (LRU) algorithm combines a traditional LRU doubly linked list with a non-blocking ring buffer to deliver high degrees of scalability on modern multicore CPUs.

- The `critnib` indexing structure delivers high performance while being very space efficient.

The cache is tuned to work optimally with relatively large value sizes. The smallest possible size is 256 bytes, but `libvmemcache` works best if the expected value sizes are above 1 kilobyte.

Chapter 10 describes this library in more detail. `libvmemcache` is an open source project on GitHub at `https://github.com/pmem/vmemcache`.

When to use it?

Use libvmemcache when implementing caching for workloads that typically would have low space efficiency when cached using a system with a normal memory allocation scheme.

libvmem

What is it?

libvmem is a deprecated predecessor to libmemkind. It is a jemalloc-derived memory allocator, with both metadata and objects allocations placed in file-based mapping. The libvmem library is an open source project available from https://pmem.io/pmdk/libvmem/.

When to use it?

Use libvmem only if you have an existing application that uses libvmem or if you need to have multiple completely separate heaps of memory. Otherwise, consider using libmemkind.

Persistent Libraries

Persistent libraries help applications maintain data structure consistency in the presence of failures. In contrast to the previously described volatile libraries, these provide new semantics and take full advantage of the unique possibilities enabled by persistent memory.

libpmem

What is it?

libpmem is a low-level C library that provides basic abstraction over the primitives exposed by the operating system. It automatically detects features available in the platform and chooses the right durability semantics and memory transfer (memcpy()) methods optimized for persistent memory. Most applications will need at least parts of this library.

Chapter 4 describes the requirements for applications using persistent memory, and Chapter 6 describes libpmem in more depth.

When to use it?

Use libpmem when modifying an existing application that already uses memory-mapped I/O. Such applications can leverage the persistent memory synchronization primitives, such as user space flushing, to replace msync(), thus reducing the kernel overhead.

Also use libpmem when you want to build everything from the ground up. It supports implementation of low-level persistent data structures with custom memory management and recovery logic.

libpmemobj

What is it?

libpmemobj is a C library that provides a transactional object store, with a manual dynamic memory allocator, transactions, and general facilities for persistent memory programming. This library solves many of the commonly encountered algorithmic and data structure problems when programming for persistent memory. Chapter 7 describes this library in detail.

When to use it?

Use libpmemobj when the programming language of choice is C and when you need flexibility in terms of data structures design but can use a general-purpose memory allocator and transactions.

libpmemobj-cpp

What is it?

libpmemobj-cpp, also known as libpmemobj++, is a C++ header-only library that uses the metaprogramming features of C++ to provide a simpler, less error-prone interface to libpmemobj. It enables rapid development of persistent memory applications by reusing many concepts C++ programmers are already familiar with, such as smart pointers and closure-based transactions.

This library also ships with custom-made, STL-compatible data structures and containers, so that application developers do not have to reinvent the basic algorithms for persistent memory.

When to use it?

When C++ is an option, `libpmemobj-cpp` is preferred for general-purpose persistent memory programming over `libpmemobj`. Chapter 7 describes this library in detail.

libpmemkv

What is it?

`libpmemkv` is a generic embedded local key-value store optimized for persistent memory. It is easy to use and ships with many different language integrations, including C, C++, and JavaScript.

This library has a pluggable back end for different storage engines. Thus, it can be used as a volatile library, although it was originally designed primarily to support persistent use cases.

Chapter 9 describes this library in detail.

When to use it?

This library is the recommended starting point into the world of persistent memory programming because it is approachable and has a simple interface. Use it when complex and custom data structures are not needed and a generic key-value store interface is enough to solve the current problem.

libpmemlog

What is it?

`libpmemlog` is a C library that implements a persistent memory append-only log file with power fail-safe operations.

When to use it?

Use `libpmemlog` when your use case exactly fits into the provided log API; otherwise, a more generic library such as `libpmemobj` or `libpmemobj-cpp` might be more useful.

libpmemblk

What is it?

`libpmemblk` is a C library for managing fixed-size arrays of blocks. It provides fail-safe interfaces to update the blocks through buffer-based functions.

When to use it?

Use libpmemblk only when a simple array of fixed blocks is needed and direct byte-level access to blocks is not required.

Tools and Command Utilities

PMDK comes with a wide variety of tools and utilities to assist in the development and deployment of persistent memory applications.

pmempool

What is it?

The pmempool utility is a tool for managing and offline analysis of persistent memory pools. Its variety of functionalities, useful throughout the entire life cycle of an application, include

- Obtaining information and statistics from a memory pool

- Checking a memory pool's consistency and repairing it if possible

- Creating memory pools

- Removing/deleting a previously created memory pool

- Updating internal metadata to the latest layout version

- Synchronizing replicas within a poolset

- Modifying internal data structures within a poolset

- Enabling or disabling pool and poolset features

When to use it?

Use pmempool whenever you are creating persistent memory pools for applications using any of the persistent libraries from PMDK.

pmemcheck

What is it?

The pmemcheck utility is a Valgrind-based tool for dynamic runtime analysis of common persistent memory errors, such as a missing flush or incorrect use of transactions. Chapter 12 describes this utility in detail.

When to use it?

The pmemcheck utility is useful when developing an application using libpmemobj, libpmemobj-cpp, or libpmem because it can help you find bugs that are common in persistent applications. We suggest running error-checking tools early in the lifetime of a codebase to avoid a pileup of hard-to-debug problems. The PMDK developers integrate pmemcheck tests into the continuous integration pipeline of PMDK, and we recommend the same for any persistent applications.

pmreorder

What is it?

The pmreorder utility helps detect data structure consistency problems of persistent applications in the presence of failures. It does this by first recording and then replaying the persistent state of the application while verifying consistency of the application's data structures at any possible intermediate state. Chapter 12 describes this utility in detail.

When to use it?

Just like pmemcheck, pmreorder is an essential tool for finding hard-to-debug persistent problems and should be integrated into the development and testing cycle of any persistent memory application.

Summary

This chapter provides a brief listing of the libraries and tools available in PMDK and when to use them. You now have enough information to know what is possible. Throughout the rest of this book, you will learn how to create software using these libraries and tools.

The next chapter introduces libpmem and describes how to use it to create simple persistent applications.

CHAPTER 6

libpmem: Low-Level Persistent Memory Support

This chapter introduces libpmem, one of the smallest libraries in PMDK. This C library is very low level, dealing with things like CPU instructions related to persistent memory, optimal ways to copy data to persistence, and file mapping. Programmers who only want completely raw access to persistent memory, without libraries to provide allocators or transactions, will likely want to use libpmem as a basis for their development.

The code in libpmem that detects the available CPU instructions, for example, is a mundane boilerplate code that you do not want to invent repeatedly in applications. Leveraging this small amount of code from libpmem will save time, and you get the benefit of fully tested and tuned code in the library.

For most programmers, libpmem is too low level, and you can safely skim this chapter quickly (or skip it altogether) and move on to the higher-level, friendlier libraries available in PMDK. All the PMDK libraries that deal with persistence, such as libpmemobj, are built on top of libpmem to meet their low-level needs.

Like all PMDK libraries, online man pages are available. For libpmem, they are at http://pmem.io/pmdk/libpmem/. This site includes links to the man pages for both the Linux and Windows version. Although the goal of the PMDK project was to make the interfaces similar across operating systems, some small differences appear as necessary. The C code examples used in this chapter build and run on both Linux and Windows.

© The Author(s) 2020
S. Scargall, *Programming Persistent Memory*, https://doi.org/10.1007/978-1-4842-4932-1_6

The examples used in this chapter are

- `simple_copy.c` is a small program that copies a 4KiB block from a source file to a destination file on persistent memory.

- `full_copy.c` is a more complete copy program, copying the entire file.

- `manpage.c` is the simple example used in the `libpmem` man page.

Using the Library

To use `libpmem`, start by including the appropriate header, as shown in Listing 6-1.

Listing 6-1. Including the libpmem headers

```
32
33  /*
34   * simple_copy.c
35   *
36   * usage: simple_copy src-file dst-file
37   *
38   * Reads 4KiB from src-file and writes it to dst-file.
39   */
40
41  #include <sys/types.h>
42  #include <sys/stat.h>
43  #include <fcntl.h>
44  #include <stdio.h>
45  #include <errno.h>
46  #include <stdlib.h>
47  #ifndef _WIN32
48  #include <unistd.h>
49  #else
50  #include <io.h>
51  #endif
52  #include <string.h>
53  #include <libpmem.h>
```

Notice the `include` on line 53. To use `libpmem`, use this include line, and link the C program with `libpmem` using the `-lpmem` option when building under Linux.

Mapping a File

The `libpmem` library contains some convenience functions for memory mapping files. Of course, your application can call `mmap()` on Linux or `MapViewOfFile()` on Windows directly, but using `libpmem` has some advantages:

- `libpmem` knows the correct arguments to the operating system mapping calls. For example, on Linux, it is not safe to flush changes to persistent memory using the CPU instructions directly unless the mapping is created with the `MAP_SYNC` flag to `mmap()`.

- `libpmem` detects if the mapping is actually persistent memory and if using the CPU instructions directly for flushing is safe.

Listing 6-2 shows how to memory map a file on a persistent memory-aware file system into the application.

Listing 6-2. Mapping a persistent memory file

```
80      /* create a pmem file and memory map it */
81      if ((pmemaddr = pmem_map_file(argv[2], BUF_LEN,
82              PMEM_FILE_CREATE|PMEM_FILE_EXCL,
83              0666, &mapped_len, &is_pmem)) == NULL) {
84          perror("pmem_map_file");
85          exit(1);
86      }
```

As part of the persistent memory detection mentioned earlier, the flag `is_pmem` is returned by `pmem_map_file`. It is the caller's responsibility to use this flag to determine how to flush changes to persistence. When making a range of memory persistent, the caller can use the optimal flush provided by `libpmem`, `pmem_persist`, only if the `is_pmem` flag is set. This is illustrated in the man page example excerpt in Listing 6-3.

Listing 6-3. manpage.c: Using the is_pmem flag

```
74      /* Flush above strcpy to persistence */
75      if (is_pmem)
76          pmem_persist(pmemaddr, mapped_len);
77      else
78          pmem_msync(pmemaddr, mapped_len);
```

Listing 6-3 shows the convenience function pmem_msync(), which is just a small wrapper around msync() or the Windows equivalent. You do not need to build in different logic for Linux and Windows because libpmem handles this.

Copying to Persistent Memory

There are several interfaces in libpmem for optimally copying or zeroing ranges of persistent memory. The simplest interface shown in Listing 6-4 is used to copy the block of data from the source file to the persistent memory in the destination file and flush it to persistence.

Listing 6-4. simple_copy.c: Copying to persistent memory

```
88      /* read up to BUF_LEN from srcfd */
89      if ((cc = read(srcfd, buf, BUF_LEN)) < 0) {
90          pmem_unmap(pmemaddr, mapped_len);
91          perror("read");
92          exit(1);
93      }
94
95      /* write it to the pmem */
96      if (is_pmem) {
97          pmem_memcpy_persist(pmemaddr, buf, cc);
98      } else {
99          memcpy(pmemaddr, buf, cc);
100         pmem_msync(pmemaddr, cc);
101     }
```

Notice how the is_pmem flag on line 96 is used just like it would be for calls to pmem_persist(), since the pmem_memcpy_persist() function includes the flush to persistence.

The interface pmem_memcpy_persist() includes the flush to persistent because it may determine that the copy is more optimally performed by using non-temporal stores, which bypass the CPU cache and do not require subsequent cache flush instructions for persistence. By providing this API, which both copies and flushes, libpmem is free to use the most optimal way to perform both steps.

Separating the Flush Steps

Flushing to persistence involves two steps:

1. Flush the CPU caches or bypass them entirely as explained in the previous example.

2. Wait for any hardware buffers to drain, to ensure writes have reached the media.

These steps are performed together when pmem_persist() is called, or they can be called individually by calling pmem_flush() for the first step and pmem_drain() for the second. Note that either of these steps may be unnecessary on a given platform, and the library knows how to check for that and do what is correct. For example, on Intel platforms, pmem_drain is an empty function.

When does it make sense to break flushing into steps? The example in Listing 6-5 illustrates one reason you might want to do this. Since the example copies data using multiple calls to memcpy(), it uses the version of libpmem copy (pmem_memcpy_nodrain()) that only performs the flush, postponing the final drain step to the end. This works because, unlike the flush step, the drain step does not take an address range; it is a system-wide drain operation so can happen at the end of the loop that copies individual blocks of data.

Listing 6-5. full_copy.c: Separating the flush steps

```
58  /*
59   * do_copy_to_pmem
60   */
61  static void
62  do_copy_to_pmem(char *pmemaddr, int srcfd, off_t len)
```

```
63  {
64      char buf[BUF_LEN];
65      int cc;
66
67      /*
68       * Copy the file,
69       * saving the last flush & drain step to the end
70       */
71      while ((cc = read(srcfd, buf, BUF_LEN)) > 0) {
72          pmem_memcpy_nodrain(pmemaddr, buf, cc);
73          pmemaddr += cc;
74      }
75
76      if (cc < 0) {
77          perror("read");
78          exit(1);
79      }
80
81      /* Perform final flush step */
82      pmem_drain();
83  }
```

In Listing 6-5, pmem_memcpy_nodrain() is specifically designed for persistent memory. When using other libraries and standard functions like memcpy(), remember they were written before persistent memory existed and do not perform any flushing to persistence. In particular, the memcpy() provided by the C runtime environment often chooses between regular stores (which require flushing) and non-temporal stores (which do not require flushing). It is making that choice based on performance, not persistence. Since you will not know which instructions it chooses, you will need to perform the flush to persistence yourself using pmem_persist() or msync().

The choice of instructions used when copying ranges to persistent memory is fairly important to the performance in many applications. The same is true when zeroing out ranges of persistent memory. To meet these needs, libpmem provides pmem_memmove(), pmem_memcpy(), and pmem_memset(), which all take a *flags* argument to give the caller more control over which instructions they use. For example, passing the flag

PMEM_F_MEM_NONTEMPORAL will tell these functions to use non-temporal stores instead of choosing which instructions to use based on the size of the range. The full list of flags is documented in the man pages for these functions.

Summary

This chapter demonstrated some of the fairly small set of APIs provided by libpmem. This library does not track what changed for you, does not provide power fail-safe transactions, and does not provide an allocator. Libraries like libpmemobj (described in the next chapter) provide all those tasks and use libpmem internally for simple flushing and copying.

libpmemobj: A Native Transactional Object Store

In the previous chapter, we described libpmem, the low-level persistent memory library that provides you with an easy way to directly access persistent memory. libpmem is a small, lightweight, and feature-limited library that is designed for software that tracks every store to pmem and needs to flush those changes to persistence. It excels at what it does. However, most developers will find higher-level libraries within the Persistent Memory Development Kit (PMDK), like libpmemobj, to be much more convenient.

This chapter describes libpmemobj, which builds upon libpmem and turns persistent memory-mapped files into a flexible object store. It supports transactions, memory management, locking, lists, and several other features.

What is libpmemobj?

The libpmemobj library provides a transactional object store in persistent memory for applications that require transactions and persistent memory management using direct access (DAX) to the memory. Briefly recapping our DAX discussion in Chapter 3, DAX allows applications to memory map files on a persistent memory-aware file system to provide direct load/store operations without paging blocks from a block storage device. It bypasses the kernel, avoids context switches and interrupts, and allows applications to read and write directly to the byte-addressable persistent storage.

© The Author(s) 2020
S. Scargall, *Programming Persistent Memory*, https://doi.org/10.1007/978-1-4842-4932-1_7

Why not malloc()?

Using libpmem seems simple. You need to flush anything you have written and use discipline when ordering such that data needs to be persisted before any pointers to it go live.

If only persistent memory programming were so simple. Apart from some specific patterns that can be done in a simpler way, such as append-only records that can be efficiently handled by libpmemlog, any new piece of data needs to have its memory allocated. When and how should the allocator mark the memory as in use? Should the allocator mark the memory as allocated before writing data or after? Neither approach works for these reasons:

- If the allocator marks the memory as allocated before the data is written, a power outage during the write can cause torn updates and a so-called "persistent leak."

- If the allocator writes the data, then marks it as allocated, a power outage that occurs between the write completing and the allocator marking it as allocated can overwrite the data when the application restarts since the allocator believes the block is available.

Another problem is that a significant number of data structures include cyclical references and thus do not form a tree. They could be implemented as a tree, but this approach is usually harder to implement.

Byte-addressable memory guarantees atomicity of only a single write. For current processors, that is generally one 64-bit word (8-bytes) that should be aligned, but this is not a requirement in practice.

All of the preceding problems could be solved if multiple writes occurred simultaneously. In the event of a power failure, any incomplete writes should either be replayed as though the power failure never happened or discarded as though the write never occurred. Applications solve this in different ways using atomic operations, transactions, redo/undo logging, etc. Using libpmemobj can solve those problems because it uses atomic transactions and redo/undo logs.

Grouping Operations

With the exception of modifying a single scalar value that fits within the processor's word, a series of data modifications must be grouped together and accompanied by a means of detecting an interruption before completion.

Memory Pools

Memory-mapped files are at the core of the persistent memory programming model. The libpmemobj library provides a convenient API to easily manage pool creation and access, avoiding the complexity of directly mapping and synchronizing data. PMDK also provides a pmempool utility to administer memory pools from the command line. Memory pools reside on DAX-mounted file systems.

Creating Memory Pools

Use the pmempool utility to create persistent memory pools for use with applications. Several pool types can be created including pmemblk, pmemlog, and pmemobj. When using libpmemobj in applications, you want to create a pool of type obj (pmemobj). Refer to the pmempool-create(1) man page for all available commands and options. The following examples are for reference:

Example 1. Create a libpmemobj (obj) type pool of minimum allowed size and layout called "my_layout" in the mounted file system /mnt/pmemfs0/

```
$ pmempool create --layout my_layout obj /mnt/pmemfs0/pool.obj
```

Example 2. Create a libpmemobj (obj) pool of 20GiB and layout called "my_layout" in the mounted file system /mnt/pmemfs0/

```
$ pmempool create --layout my_layout --size 20G obj \
/mnt/pmemfs0/pool.obj
```

Example 3. Create a `libpmemobj` (obj) pool using all available capacity within the /mnt/pmemfs0/ file system using the layout name of "my_layout"

```
$ pmempool create --layout my_layout --max-size obj \
/mnt/pmemfs0/pool.obj
```

Applications can programmatically create pools that do not exist at application start time using `pmemobj_create()`. `pmemobj_create()` has the following arguments:

```
PMEMobjpool *pmemobj_create(const char *path,
    const char *layout, size_t poolsize, mode_t mode);
```

- `path` specifies the name of the memory pool file to be created, including a full or relative path to the file.

- `layout` specifies the application's layout type in the form of a string to identify the pool.

- `poolsize` specifies the required size for the pool. The memory pool file is fully allocated to the size `poolsize` using `posix_fallocate(3)`. The minimum size for a pool is defined as `PMEMOBJ_MIN_POOL` in `<libpmemobj.h>`. If the pool already exists, `pmemobj_create()` will return an `EEXISTS` error. Specifying `poolsize` as zero will take the pool size from the file size and will verify that the file appears to be empty by searching for any nonzero data in the pool header at the beginning of the file.

- `mode` specifies the ACL permissions to use when creating the file, as described by `create(2)`.

Listing 7-1 shows how to create a pool using the `pmemobj_create()` function.

Listing 7-1. pwriter.c – An example showing how to create a pool using pmemobj_create()

```
33   /*
34    * pwriter.c -  Write a string to a
35    *             persistent memory pool
36    */
37
```

```
38  #include <stdio.h>
39  #include <string.h>
40  #include <libpmemobj.h>
41
42  #define LAYOUT_NAME "rweg"
43  #define MAX_BUF_LEN 31
44
45  struct my_root {
46      size_t len;
47      char buf[MAX_BUF_LEN];
48  };
49
50  int
51  main(int argc, char *argv[])
52  {
53      if (argc != 2) {
54          printf("usage: %s file-name\n", argv[0]);
55          return 1;
56      }
57
58      PMEMobjpool *pop = pmemobj_create(argv[1],
59          LAYOUT_NAME, PMEMOBJ_MIN_POOL, 0666);
60
61      if (pop == NULL) {
62          perror("pmemobj_create");
63          return 1;
64      }
65
66      PMEMoid root = pmemobj_root(pop,
67          sizeof(struct my_root));
68
69      struct my_root *rootp = pmemobj_direct(root);
70
71      char buf[MAX_BUF_LEN] = "Hello PMEM World";
72
```

```
73      rootp->len = strlen(buf);
74      pmemobj_persist(pop, &rootp->len,
75          sizeof(rootp->len));
76
77      pmemobj_memcpy_persist(pop, rootp->buf, buf,
78          rootp->len);
79
80      pmemobj_close(pop);
81
82      return 0;
83  }
```

- Line 42: We define the name for our pool layout to be "rweg" (read-write example). This is just a name and can be any string that uniquely identifies the pool to the application. A NULL value is valid. In the case where multiple pools are opened by the application, this name uniquely identifies it.

- Line 43: We define the maximum length of the write buffer.

- Lines 45-47: This defines the root object data structure which has members len and buf. buf contains the string we want to write, and the len is the length of the buffer.

- Lines 53- 56: The pwriter command accepts one argument: the path and pool name to write to. For example, /mnt/pmemfs0/helloworld_obj.pool. The file name extension is arbitrary and optional.

- Lines 58-59: We call pmemobj_create() to create the pool using the file name passed in from the command line, the layout name of "rweg," a size we set to be the minimum size for an object pool type, and permissions of 0666. We cannot create a pool smaller than defined by PMEMOBJ_MIN_POOL or larger than the available space on the file system. Since the string in our example is very small, we only require a minimally sized pool. On success, pmemobj_create() returns a pool object pointer (POP) of type PMEMobjpool, that we can use to acquire a pointer to the root object.

- Lines 61-64: If `pmemobj_create()` fails, we will exit the program and return an error.

- Line 66: Using the pop acquired from line 58, we use the `pmemobj_root()` function to locate the `root` object.

- Line 69: We use the `pmemobj_direct()` function to get a pointer to the root object we found in line 66.

- Line 71: We set the string/buffer to "Hello PMEM World."

- Lines 73-78. After determining the length of the buffer, we first write the `len` and then the `buf` member of our `root` object to persistent memory.

- Line 80: We close the persistent memory pool by unmapping it.

Pool Object Pointer (POP) and the Root Object

Due to the address space layout randomization (ASLR) feature used by most operating systems, the location of the pool – once memory mapped into the application address space – can differ between executions and system reboots. Without a way to access the data within the pool, you would find it challenging to locate the data within a pool. PMDK-based pools have a small amount of metadata to solve this problem.

Every `pmemobj` (obj) type pool has a `root` object. This `root` object is necessary because it is used as an entry point from which to find all the other objects created in a pool, that is, user data. An application will locate the `root` object using a special object called pool object pointer (POP). The POP object resides in volatile memory and is created with every program invocation. It keeps track of metadata related to the pool, such as the offset to the root object inside the pool. Figure 7-1 depicts the POP and memory pool layout.

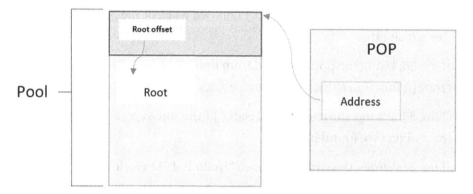

Figure 7-1. *A high-level overview of a persistent memory pool with a pool object pointer (POP) pointing to the root object*

Using a valid pop pointer, you can use the pmemobj_root() function to get a pointer of the root object. Internally, this function creates a valid pointer by adding the current memory address of the mapped pool plus the internal offset to the root.

Opening and Reading from Memory Pools

You create a pool using pmemobj_create(), and you open an existing pool using pmemobj_open(). Both functions return a PMEMobjpool *pop pointer. The pwriter example in Listing 7-1 shows how to create a pool and write a string to it. Listing 7-2 shows how to open the same pool to read and display the string.

Listing 7-2. preader.c – An example showing how to open a pool and access the root object and data

```
33  /*
34   * preader.c -  Read a string from a
35   *              persistent memory pool
36   */
37
38  #include <stdio.h>
39  #include <string.h>
40  #include <libpmemobj.h>
41
```

```
42  #define LAYOUT_NAME "rweg"
43  #define MAX_BUF_LEN 31
44
45  struct my_root {
46      size_t len;
47      char buf[MAX_BUF_LEN];
48  };
49
50  int
51  main(int argc, char *argv[])
52  {
53      if (argc != 2) {
54          printf("usage: %s file-name\n", argv[0]);
55          return 1;
56      }
57
58      PMEMobjpool *pop = pmemobj_open(argv[1],
59          LAYOUT_NAME);
60
61      if (pop == NULL) {
62          perror("pmemobj_open");
63          return 1;
64      }
65
66      PMEMoid root = pmemobj_root(pop,
67          sizeof(struct my_root));
68      struct my_root *rootp = pmemobj_direct(root);
69
70      if (rootp->len == strlen(rootp->buf))
71          printf("%s\n", rootp->buf);
72
73      pmemobj_close(pop);
74
75      return 0;
76  }
```

- Lines 42-48: We use the same data structure declared in pwriter.c. In practice, this should be declared in a header file for consistency.

- Line 58: Open the pool and return a pop pointer to it

- Line 66: Upon success, pmemobj_root() returns a handle to the root object associated with the persistent memory pool pop.

- Line 68: pmemobj_direct() returns a pointer to the root object.

- Lines 70-71: Determine the length of the buffer pointed to by rootp->buf. If it matches the length of the buffer we wrote, the contents of the buffer is printed to STDOUT.

Memory Poolsets

The capacity of multiple pools can be combined into a *poolset*. Besides providing a way to increase the available space, a poolset can be used to span multiple persistent memory devices and provide both local and remote replication.

You open a poolset the same way as a single pool using pmemobj_open(). (At the time of publication, pmemobj_create() and the pmempool utility cannot create poolsets. Enhancement requests exist for these features.) Although creating poolsets requires manual administration, poolset management can be automated via libpmempool or the pmempool utility; full details appear in the poolset(5) man page.

Concatenated Poolsets

Individual pools can be concatenated using pools on a single or multiple file systems. Concatenation only works with the same pool type: block, object, or log pools. Listing 7-3 shows an example "myconcatpool.set" poolset file that concatenates three smaller pools into a larger pool. For illustrative purposes, each pool is a different size and located on different file systems. An application using this poolset would see a single 700GiB memory pool.

Listing 7-3. myconcatpool.set – An example of a concatenated poolset created from three individual pools on three different file systems

```
PMEMPOOLSET
OPTION NOHDRS
100G /mountpoint0/myfile.part0
200G /mountpoint1/myfile.part1
400G /mountpoint2/myfile.part2
```

Note   Data will be preserved if it exists in /mountpoint0/myfile.part0, but any data in /mountpoint0/myfile.part1 or /mountpoint0/myfile.part2 will be lost. We recommend that you only add new and empty pools to a poolset.

Replica Poolsets

Besides combining multiple pools to provide more space, a poolset can also maintain multiple copies of the same data to increase resiliency. Data can be replicated to another poolset on a different file of the local host and a poolset on a remote host.

Listing 7-4 shows a poolset file called "myreplicatedpool.set" that will replicate local writes into the /mnt/pmem0/pool1 pool to another local pool, /mnt/pmem1/pool1, on a different file system, and to a remote-objpool.set poolset on a remote host called example.com.

Listing 7-4. myreplicatedpool.set – An example demonstrating how to replicate local data locally and remote host

```
PMEMPOOLSET
256G /mnt/pmem0/pool1

REPLICA
256G /mnt/pmem1/pool1

REPLICA user@example.com remote-objpool.set
```

The librpmem library, a remote persistent memory support library, underpins this feature. Chapter 18 discusses librpmem and replica pools in more detail.

Managing Memory Pools and Poolsets

The pmempool utility has several features that developers and system administrators may find useful. We do not present their details here because each command has a detailed man page:

- **pmempool info** prints information and statistics in human-readable format about the specified pool.

- **pmempool check** checks the pool's consistency and repairs pool if it is not consistent.

- **pmempool create** creates a pool of specified type with additional properties specific for this type of pool.

- **pmempool dump** dumps usable data from a pool in hexadecimal or binary format.

- **pmempool rm** removes pool file or all pool files listed in pool set configuration file.

- **pmempool convert** updates the pool to the latest available layout version.

- **pmempool sync** synchronizes replicas within a poolset.

- **pmempool transform** modifies the internal structure of a poolset.

- **pmempool feature** toggles or queries a poolset's features.

Typed Object Identifiers (TOIDs)

When we write data to a persistent memory pool or device, we commit it at a physical address. With the ASLR feature of operating systems, when applications open a pool and memory map it into the address space, the virtual address will change each time. For this reason, a type of handle (pointer) that does not change is needed; this handle is called an OID (object identifier). Internally, it is a pair of the pool or poolset unique identifier (UUID) and an offset within the pool or poolset. The OID can be translated back and forth between its persistent form and pointers that are fit for direct use by this particular instance of your program.

At a low level, the translation can be done manually via functions such as pmemobj_direct() that appear in the preader.c example in Listing 7-2. Because manual translations require explicit type casts and are error prone, we recommend tagging every object with a type. This allows some form of type safety, and thanks to macros, can be checked at compile time.

For example, a persistent variable declared via TOID(struct foo) x can be read via D_RO(x)->field. In a pool with the following layout:

```
POBJ_LAYOUT_BEGIN(cathouse);
POBJ_LAYOUT_TOID(cathouse, struct canaries);
POBJ_LAYOUT_TOID(cathouse, int);
POBJ_LAYOUT_END(cathouse);
```

The field val declared on the first line can be accessed using any of the subsequent three operations:

```
TOID(int) val;
TOID_ASSIGN(val, oid_of_val); // Assigns 'oid_of_val' to typed OID 'val'
D_RW(val) = 42; // Returns a typed write pointer to 'val' and writes 42
return D_RO(val); // Returns a typed read-only (const) pointer to 'val'
```

Allocating Memory

Using malloc() to allocate memory is quite normal to C developers and those who use languages that do not fully handle automatic memory allocation and deallocation. For persistent memory, you can use pmemobj_alloc(), pmemobj_reserve(), or pmemobj_xreserve() to reserve memory for a transient object and use it the same way you would use malloc(). We recommend that you free allocated memory using pmemobj_free() or POBJ_FREE() when the application no longer requires it to avoid a runtime memory leak. Because these are volatile memory allocations, they will not cause a persistent leak after a crash or graceful application exit.

Persisting Data

The typical intent of using persistent memory is to save data persistently. For this, you need to use one of three APIs that `libpmemobj` provides:

- Atomic operations
- Reserve/publish
- Transactional

Atomic Operations

The `pmemobj_alloc()` and its variants shown below are easy to use, but they are limited in features, so additional coding is required by the developer:

```
int pmemobj_alloc(PMEMobjpool *pop, PMEMoid *oidp,
    size_t size, uint64_t type_num, pmemobj_constr
    constructor, void *arg);
int pmemobj_zalloc(PMEMobjpool *pop, PMEMoid *oidp,
    size_t size, uint64_t type_num);
void pmemobj_free(PMEMoid *oidp);
int pmemobj_realloc(PMEMobjpool *pop, PMEMoid *oidp,
    size_t size, uint64_t type_num);
int pmemobj_zrealloc(PMEMobjpool *pop, PMEMoid *oidp,
    size_t size, uint64_t type_num);
int pmemobj_strdup(PMEMobjpool *pop, PMEMoid *oidp,
    const char *s, uint64_t type_num);
int pmemobj_wcsdup(PMEMobjpool *pop, PMEMoid *oidp,
    const wchar_t *s, uint64_t type_num);
```

The TOID-based wrappers for most of these functions include:

```
POBJ_NEW(PMEMobjpool *pop, TOID *oidp, TYPE,
    pmemobj_constr constructor, void *arg)
POBJ_ALLOC(PMEMobjpool *pop, TOID *oidp, TYPE, size_t size,
    pmemobj_constr constructor, void *arg)
POBJ_ZNEW(PMEMobjpool *pop, TOID *oidp, TYPE)
POBJ_ZALLOC(PMEMobjpool *pop, TOID *oidp, TYPE, size_t size)
```

```
POBJ_REALLOC(PMEMobjpool *pop, TOID *oidp, TYPE, size_t size)
POBJ_ZREALLOC(PMEMobjpool *pop, TOID *oidp, TYPE, size_t size)
POBJ_FREE(TOID *oidp)
```

These functions reserve the object in a temporary state, call the constructor you provided, and then in one atomic action, mark the allocation as persistent. They will insert the pointer to the newly initialized object into a variable of your choice.

If the new object needs to be merely zeroed, pmemobj_zalloc() does so without requiring a constructor.

Because copying NULL-terminated strings is a common operation, libpmemobj provides pmemobj_strdup() and its wide-char variant pmemobj_wcsdup() to handle this. pmemobj_strdup() provides the same semantics as strdup(3) but operates on the persistent memory heap associated with the memory pool.

Once you are done with the object, pmemobj_free() will deallocate the object while zeroing the variable that stored the pointer to it. The pmemobj_free() function frees the memory space represented by oidp, which must have been allocated by a previous call to pmemobj_alloc(), pmemobj_xalloc(), pmemobj_zalloc(), pmemobj_realloc(), or pmemobj_zrealloc(). The pmemobj_free() function provides the same semantics as free(3), but instead of operating on the process heap supplied by the system, it operates on the persistent memory heap.

Listing 7-5 shows a small example of allocating and freeing memory using the libpmemobj API.

Listing 7-5. Using pmemobj_alloc() to allocate memory and using pmemobj_free() to free it

```
33  /*
34   * pmemobj_alloc.c - An example to show how to use
35   *                   pmemobj_alloc()
36   */
..
47  typedef uint32_t color;
48
49  static int paintball_init(PMEMobjpool *pop,
50          void *ptr, void *arg)
51  {
52      *(color *)ptr = time(0) & 0xffffff;
```

```
53        pmemobj_persist(pop, ptr, sizeof(color));
54        return 0;
55    }
56
57    int main()
58    {
59        PMEMobjpool *pool = pmemobj_open(POOL, LAYOUT);
60        if (!pool) {
61            pool = pmemobj_create(POOL, LAYOUT,
62            PMEMOBJ_MIN_POOL, 0666);
63            if (!pool)
64                die("Couldn't open pool: %m\n");
65
66        }
67        PMEMoid root = pmemobj_root(pool,
68                sizeof(PMEMoid) * 6);
69        if (OID_IS_NULL(root))
70            die("Couldn't access root object.\n");
71
72        PMEMoid *chamber = (PMEMoid *)pmemobj_direct(root)
73            + (getpid() % 6);
74        if (OID_IS_NULL(*chamber)) {
75            printf("Reloading.\n");
76            if (pmemobj_alloc(pool, chamber, sizeof(color)
77                , 0, paintball_init, 0))
78                die("Failed to alloc: %m\n");
79        } else {
80            printf("Shooting %06x colored bullet.\n",
81            *(color *)pmemobj_direct(*chamber));
82            pmemobj_free(chamber);
83        }
84
85        pmemobj_close(pool);
86        return 0;
87    }
```

- Line 47: Defines a color that will be stored in the pool.

- Lines 49-54: The `paintball_init()` function is called when we allocate memory (line 76). This function takes a pool and object pointer, calculates a random hex value for the paintball color, and persistently writes it to the pool. The program exits when the write completes.

- Lines 59-70: Opens or creates a pool and acquires a pointer to the root object within the pool.

- Line 72: Obtain a pointer to an offset within the pool.

- Lines 74-78: If the pointer in line 72 is not a valid object, we allocate some space and call `paintball_init()`.

- Lines 79-80: If the pointer in line 72 is a valid object, we read the color value, print the string, and free the object.

Reserve/Publish API

The atomic allocation API will not help if

- There is more than one reference to the object that needs to be updated

- There are multiple scalars that need to be updated

For example, if your program needs to subtract money from account A and add it to account B, both operations must be done together. This can be done via the reserve/publish API.

To use it, you specify any number of operations to be done. The operations may be setting a scalar 64-bit value using `pmemobj_set_value()`, freeing an object with `pmemobj_defer_free()`, or allocating it using `pmemobj_reserve()`. Of these, only the allocation happens immediately, letting you do any initialization of the newly reserved object. Modifications will not become persistent until `pmemobj_publish()` is called.

Functions provided by `libpmemobj` related to the reserve/publish feature are

```
PMEMoid pmemobj_reserve(PMEMobjpool *pop,
    struct pobj_action *act, size_t size, uint64_t type_num);
void pmemobj_defer_free(PMEMobjpool *pop, PMEMoid oid,
```

```
    struct pobj_action *act);
void pmemobj_set_value(PMEMobjpool *pop,
    struct pobj_action *act, uint64_t *ptr, uint64_t value);
int pmemobj_publish(PMEMobjpool *pop,
    struct pobj_action *actv, size_t actvcnt);
void pmemobj_cancel(PMEMobjpool *pop,
    struct pobj_action *actv, size_t actvcnt);
```

Listing 7-6 is a simple banking example that demonstrates how to change multiple scalars (account balances) before publishing the updates into the pool.

Listing 7-6. Using the reserve/publish API to modify bank account balances

```
32
33  /*
34   * reserve_publish.c - An example using the
35   *                      reserve/publish libpmemobj API
36   */
37

..
44  #define POOL "/mnt/pmem/balance"
45
46  static PMEMobjpool *pool;
47
48  struct account {
49      PMEMoid name;
50      uint64_t balance;
51  };
52  TOID_DECLARE(struct account, 0);
53

..
60  static PMEMoid new_account(const char *name,
61                  int deposit)
62  {
63      int len = strlen(name) + 1;
64
65      struct pobj_action act[2];
```

```
66      PMEMoid str = pmemobj_reserve(pool, act + 0,
67                      len, 0);
68      if (OID_IS_NULL(str))
69          die("Can't allocate string: %m\n");
..
75      pmemobj_memcpy(pool, pmemobj_direct(str), name,
76                      len, PMEMOBJ_F_MEM_NODRAIN);
77      TOID(struct account) acc;
78      PMEMoid acc_oid = pmemobj_reserve(pool, act + 1,
79                          sizeof(struct account), 1);
80      TOID_ASSIGN(acc, acc_oid);
81      if (TOID_IS_NULL(acc))
82          die("Can't allocate account: %m\n");
83      D_RW(acc)->name = str;
84      D_RW(acc)->balance = deposit;
85      pmemobj_persist(pool, D_RW(acc),
86                      sizeof(struct account));
87      pmemobj_publish(pool, act, 2);
88      return acc_oid;
89  }
90
91  int main()
92  {
93      if (!(pool = pmemobj_create(POOL, "",
94                          PMEMOBJ_MIN_POOL, 0600)))
95          die("Can't create pool \"%s\": %m\n", POOL);
96
97      TOID(struct account) account_a, account_b;
98      TOID_ASSIGN(account_a,
99                  new_account("Julius Caesar", 100));
100     TOID_ASSIGN(account_b,
101                 new_account("Mark Anthony", 50));
102
103     int price = 42;
104     struct pobj_action act[2];
```

```
105        pmemobj_set_value(pool, &act[0],
106                        &D_RW(account_a)->balance,
107                        D_RW(account_a)->balance - price);
108        pmemobj_set_value(pool, &act[1],
109                        &D_RW(account_b)->balance,
110                        D_RW(account_b)->balance + price);
111        pmemobj_publish(pool, act, 2);
112
113        pmemobj_close(pool);
114        return 0;
115    }
```

- Line 44: Defines the location of the memory pool.

- Lines 48-52: Declares an account data structure with a name and balance.

- Lines 60-89: The new_account() function reserves the memory (lines 66 and 78), updates the name and balance (lines 83 and 84), persists the changes (line 85), and then publishes the updates (line 87).

- Lines 93-95: Create a new pool or exit on failure.

- Line 97: Declare two account instances.

- Lines 98-101: Create a new account for each owner with initial balances.

- Lines 103-111: We subtract 42 from Julius Caesar's account and add 42 to Mark Anthony's account. The modifications are published on line 111.

Transactional API

The reserve/publish API is fast, but it does not allow reading data you have just written. In such cases, you can use the transactional API.

The first time a variable is written, it must be explicitly added to the transaction. This can be done via pmemobj_tx_add_range() or its variants (xadd, _direct). Convenient macros such as TX_ADD() or TX_SET() can perform the same operation. The transaction-based functions and macros provided by libpmemobj include

```
int pmemobj_tx_add_range(PMEMoid oid, uint64_t off,
    size_t size);
int pmemobj_tx_add_range_direct(const void *ptr, size_t size);

TX_ADD(TOID o)
TX_ADD_FIELD(TOID o, FIELD)
TX_ADD_DIRECT(TYPE *p)
TX_ADD_FIELD_DIRECT(TYPE *p, FIELD)

TX_SET(TOID o, FIELD, VALUE)
TX_SET_DIRECT(TYPE *p, FIELD, VALUE)
TX_MEMCPY(void *dest, const void *src, size_t num)
TX_MEMSET(void *dest, int c, size_t num)
```

The transaction may also allocate entirely new objects, reserve their memory, and then persistently allocate them only one transaction commit. These functions include

```
PMEMoid pmemobj_tx_alloc(size_t size, uint64_t type_num);
PMEMoid pmemobj_tx_zalloc(size_t size, uint64_t type_num);
PMEMoid pmemobj_tx_realloc(PMEMoid oid, size_t size,
    uint64_t type_num);
PMEMoid pmemobj_tx_zrealloc(PMEMoid oid, size_t size,
    uint64_t type_num);
PMEMoid pmemobj_tx_strdup(const char *s, uint64_t type_num);
PMEMoid pmemobj_tx_wcsdup(const wchar_t *s,
    uint64_t type_num);
```

We can rewrite the banking example from Listing 7-6 using the transaction API. Most of the code remains the same except when we want to add or subtract amounts from the balance; we encapsulate those updates in a transaction, as shown in Listing 7-7.

Listing 7-7. Using the transaction API to modify bank account balances

```
33  /*
34   * tx.c - An example using the transaction API
35   */
36

..
```

```
 94   int main()
 95   {
 96       if (!(pool = pmemobj_create(POOL, "",
 97                           PMEMOBJ_MIN_POOL, 0600)))
 98           die("Can't create pool "%s": %m\n", POOL);
 99
100       TOID(struct account) account_a, account_b;
101       TOID_ASSIGN(account_a,
102                       new_account("Julius Caesar", 100));
103       TOID_ASSIGN(account_b,
104                       new_account("Mark Anthony", 50));
105
106       int price = 42;
107       TX_BEGIN(pool) {
108           TX_ADD_DIRECT(&D_RW(account_a)->balance);
109           TX_ADD_DIRECT(&D_RW(account_b)->balance);
110           D_RW(account_a)->balance -= price;
111           D_RW(account_b)->balance += price;
112       } TX_END
113
114       pmemobj_close(pool);
115       return 0;
116   }
```

- Line 107: We start the transaction.

- Lines 108-111: Make balance modifications to multiple accounts.

- Line 112: Finish the transaction. All updates will either complete
  entirely or they will be rolled back if the application or system crashes
  before the transaction completes.

Each transaction has multiple stages in which an application can interact. These
transaction stages include

- TX_STAGE_NONE: No open transaction in this thread.

- TX_STAGE_WORK: Transaction in progress.

- TX_STAGE_ONCOMMIT: Successfully committed.

- TX_STAGE_ONABORT: The transaction start either failed or was aborted.

- TX_STAGE_FINALLY: Ready for cleanup.

The example in Listing 7-7 uses the two mandatory stages: TX_BEGIN and TX_END. However, we could easily have added the other stages to perform actions for the other stages, for example:

```
TX_BEGIN(Pop) {
        /* the actual transaction code goes here... */
} TX_ONCOMMIT {
        /*
         * optional - executed only if the above block
         * successfully completes
         */
} TX_ONABORT {
        /*
         * optional - executed only if starting the transaction
         * fails, or if transaction is aborted by an error or a
         * call to pmemobj_tx_abort()
         */
} TX_FINALLY {
        /*
         * optional - if exists, it is executed after
         * TX_ONCOMMIT or TX_ONABORT block
         */
} TX_END /* mandatory */
```

Optionally, you can provide a list of parameters for the transaction. Each parameter consists of a type followed by one of these type-specific number of values:

- TX_PARAM_NONE is used as a termination marker with no following value.

- TX_PARAM_MUTEX is followed by one value, a pmem-resident PMEMmutex.

- TX_PARAM_RWLOCK is followed by one value, a pmem-resident PMEMrwlock.

- TX_PARAM_CB is followed by two values: a callback function of type pmemobj_tx_callback and a void pointer.

Using TX_PARAM_MUTEX or TX_PARAM_RWLOCK causes the specified lock to be acquired at the beginning of the transaction. TX_PARAM_RWLOCK acquires the lock for writing. It is guaranteed that pmemobj_tx_begin() will acquire all locks prior to successful completion, and they will be held by the current thread until the outermost transaction is finished. Locks are taken in order from left to right. To avoid deadlocks, you are responsible for proper lock ordering.

TX_PARAM_CB registers the specified callback function to be executed at each transaction stage. For TX_STAGE_WORK, the callback is executed prior to commit. For all other stages, the callback is executed as the first operation after a stage change. It will also be called after each transaction.

Optional Flags

Many of the functions discussed for the atomic, reserve/publish, and transactional APIs have a variant with a "flags" argument that accepts these values:

- POBJ_XALLOC_ZERO zeroes the object allocated.

- POBJ_XALLOC_NO_FLUSH suppresses automatic flushing. It is expected that you flush the data in some way; otherwise, it may not be durable in case of an unexpected power loss.

Persisting Data Summary

The atomic, reserve/publish, and transactional APIs have different strengths:

- Atomic allocations are the simplest and fastest, but their use is limited to allocating and initializing wholly new blocks.

- The reserve/publish API can be as fast as atomic allocations when all operations involve either allocating or deallocating whole objects or modifying scalar values. However, being able to read the data you have just written may be desirable.

- The transactional API requires slow synchronization whenever a variable is added to the transaction. If the variable is changed multiple times during the transaction, subsequent operations are free. It also allows conveniently mutating pieces of data larger than a single machine word.

Guarantees of libpmemobj's APIs

The transactional, atomic allocation, and reserve/publish APIs within `libpmemobj` all provide fail-safe atomicity and consistency.

The transactional API ensures the durability of any modifications of memory for an object that has been added to the transaction. An exception is when the `POBJ_X***_NO_FLUSH` flag is used, in which case the application is responsible for either flushing that memory range itself or using the `memcpy`-like functions from `libpmemobj`. The no-flush flag does not provide any isolation between threads, meaning partial writes are immediately visible to other threads.

The atomic allocation API requires that applications flush the writes done by the object's constructor. This ensures durability if the operation succeeded. It is the only API that provides full isolation between threads.

The reserve/publish API requires explicit flushes of writes to memory blocks allocated via `pmemobj_reserve()` that will flush writes done via `pmemobj_set_value()`. There is no isolation between threads, although no modifications go live until `pmemobj_publish()` starts, allowing you to take explicit locks for just the publishing stage.

Using terms known from databases, the isolation levels provided are

- Transactional API: `READ_UNCOMMITTED`

- Atomic allocations API: `READ_COMMITTED`

- Reserve/publish API: `READ_COMMITTED` until publishing starts, then `READ_UNCOMMITTED`

Managing Library Behavior

The pmemobj_set_funcs() function allows an application to override memory allocation calls used internally by libpmemobj. Passing in NULL for any of the handlers will cause the libpmemobj default function to be used. The library does not make heavy use of the system malloc() functions, but it does allocate approximately 4–8 kilobytes for each memory pool in use.

By default, libpmemobj supports up to 1024 parallel transactions/allocations. For debugging purposes, it is possible to decrease this value by setting the PMEMOBJ_NLANES shell environment variable to the desired limit. For example, at the shell prompt, run "export PMEMOBJ_NLANES=512" then run the application:

```
$ export PMEMOBJ_NLANES=512
$ ./my_app
```

To return to the default behavior, unset PMEMOBJ_NLANES using

```
$ unset PMEMOBJ_NLANES
```

Debugging and Error Handling

If an error is detected during the call to a libpmemobj function, the application may retrieve an error message describing the reason for the failure from pmemobj_errormsg(). This function returns a pointer to a static buffer containing the last error message logged for the current thread. If errno was set, the error message may include a description of the corresponding error code as returned by strerror(3). The error message buffer is thread local; errors encountered in one thread do not affect its value in other threads. The buffer is never cleared by any library function; its content is significant only when the return value of the immediately preceding call to a libpmemobj function indicated an error, or if errno was set. The application must not modify or free the error message string, but it may be modified by subsequent calls to other library functions.

Two versions of libpmemobj are typically available on a development system. The non-debug version is optimized for performance and used when a program is linked using the -lpmemobj option. This library skips checks that impact performance, never logs any trace information, and does not perform any runtime assertions.

A debug version of libpmemobj is provided and available in /usr/lib/pmdk_debug or /usr/local/lib64/pmdk_debug. The debug version contains runtime assertions and tracepoints.

The common way to use the debug version is to set the environment variable LD_LIBRARY_PATH. Alternatively, you can use LD_PRELOAD to point to /usr/lib/pmdk_debug or /usr/lib64/pmdk_debug, as appropriate. These libraries may reside in a different location, such as /usr/local/lib/pmdk_debug and /usr/local/lib64/pmdk_debug, depending on your Linux distribution or if you compiled installed PMDK from source code and chose /usr/local as the installation path. The following examples are equivalent methods for loading and using the debug versions of libpmemobj with an application called my_app:

```
$ export LD_LIBRARY_PATH=/usr/lib64/pmdk_debug
$ ./my_app
```

Or

```
$ LD_PRELOAD=/usr/lib64/pmdk_debug ./my_app
```

The output provided by the debug library is controlled using the PMEMOBJ_LOG_LEVEL and PMEMOBJ_LOG_FILE environment variables. These variables have no effect on the non-debug version of the library.

PMEMOBJ_LOG_LEVEL

The value of PMEMOBJ_LOG_LEVEL enables tracepoints in the debug version of the library, as follows:

1. This is the default level when PMEMOBJ_LOG_LEVEL is not set. No log messages are emitted at this level.

2. Additional details on any errors detected are logged, in addition to returning the errno-based errors as usual. The same information may be retrieved using pmemobj_errormsg().

3. A trace of basic operations is logged.

4. Enables an extensive amount of function-call tracing in the library.

5. Enables voluminous and fairly obscure tracing information that is likely only useful to the libpmemobj developers.

Debug output is written to STDERR unless PMEMOBJ_LOG_FILE is set. To set a debug level, use

```
$ export PMEMOBJ_LOG_LEVEL=2
$ ./my_app
```

PMEMOBJ_LOG_FILE

The value of PMEMOBJ_LOG_FILE includes the full path and file name of a file where all logging information should be written. If PMEMOBJ_LOG_FILE is not set, logging output is written to STDERR.

The following example defines the location of the log file to /var/tmp/libpmemobj_debug.log, ensures we are using the debug version of libpmemobj when executing my_app in the background, sets the debug log level to 2, and monitors the log in real time using tail -f:

```
$ export PMEMOBJ_LOG_FILE=/var/tmp/libpmemobj_debug.log
$ export PMEMOBJ_LOG_LEVEL=2
$ LD_PRELOAD=/usr/lib64/pmdk_debug ./my_app &
$ tail -f /var/tmp/libpmemobj_debug.log
```

If the last character in the debug log file name is "-", the process identifier (PID) of the current process will be appended to the file name when the log file is created. This is useful if you are debugging multiple processes.

Summary

This chapter describes the libpmemobj library, which is designed to simplify persistent memory programming. By providing APIs that deliver atomic operations, transactions, and reserve/publish features, it makes creating applications less error prone while delivering guarantees for data integrity.

libpmemobj-cpp: The Adaptable Language - C++ and Persistent Memory

Introduction

The Persistent Memory Development Kit (PMDK) includes several separate libraries; each is designed with a specific use in mind. The most flexible and powerful one is libpmemobj. It complies with the persistent memory programming model without modifying the compiler. Intended for developers of low-level system software and language creators, the libpmemobj library provides allocators, transactions, and a way to automatically manipulate objects. Because it does not modify the compiler, its API is verbose and macro heavy.

To make persistent memory programming easier and less error prone, higher-level language bindings for libpmemobj were created and included in PMDK. The C++ language was chosen to create new and friendly API to libpmemobj called libpmemobj-cpp, which is also referred to as libpmemobj++. C++ is versatile, feature rich, has a large developer base, and it is constantly being improved with updates to the C++ programming standard.

The main goal for the libpmemobj-cpp bindings design was to focus modifications to volatile programs on data structures and not on the code. In other words, libpmemobj-cpp bindings are for developers, who want to modify volatile applications, provided with a convenient API for modifying structures and classes with only slight modifications to functions.

111

S. Scargall, *Programming Persistent Memory*, https://doi.org/10.1007/978-1-4842-4932-1_8

This chapter describes how to leverage the C++ language features that support metaprogramming to make persistent memory programming easier. It also describes how to make it more C++ idiomatic by providing persistent containers. Finally, we discuss C++ standard limitations for persistent memory programming, including an object's lifetime and the internal layout of objects stored in persistent memory.

Metaprogramming to the Rescue

Metaprogramming is a technique in which computer programs have the ability to treat other programs as their data. It means that a program can be designed to read, generate, analyze or transform other programs, and even modify itself while running. In some cases, this allows programmers to minimize the number of lines of code to express a solution, in turn reducing development time. It also allows programs greater flexibility to efficiently handle new situations without recompilation.

For the `libpmemobj-cpp` library, considerable effort was put into encapsulating the PMEMoids (persistent memory object IDs) with a type-safe container. Instead of a sophisticated set of macros for providing type safety, templates and metaprogramming are used. This significantly simplifies the native C `libpmemobj` API.

Persistent Pointers

The persistent memory programming model created by the Storage Networking Industry Association (SNIA) is based on memory-mapped files. PMDK uses this model for its architecture and design implementation. We discussed the SNIA programming model in Chapter 3.

Most operating systems implement address space layout randomization (ASLR). ASLR is a computer security technique involved in preventing exploitation of memory corruption vulnerabilities. To prevent an attacker from reliably jumping to, for example, a particular exploited function in memory, ASLR randomly arranges the address space positions of key data areas of a process, including the base of the executable and the positions of the stack, heap, and libraries. Because of ASLR, files can be mapped at different addresses of the process address space each time the application executes. As a result, traditional pointers that store absolute addresses cannot be used. Upon each execution, a traditional pointer might point to uninitialized memory for which dereferencing it may result in a segmentation fault. Or it might point to a valid memory range, but not the one that the user expects it to point to, resulting in unexpected and undetermined behavior.

To solve this problem in persistent memory programming, a different type of pointer is needed. `libpmemobj` introduced a C struct called `PMEMoid`, which consists of an identifier of the pool and an offset from its beginning. This *fat pointer* is encapsulated in `libpmemobj` C++ bindings as a template class `pmem::obj::persistent_ptr`. Both the C and C++ implementations have the same 16-byte footprint. A constructor from raw `PMEMoid` is provided so that mixing the C API with C++ is possible. The `pmem::obj::persistent_ptr` is similar in concept and implementation to the smart pointers introduced in C++11 (`std::shared_ptr`, `std::auto_ptr`, `std::unique_ptr`, and `std::weak_ptr`), with one big difference – it does not manage the object's life cycle.

Besides `operator*`, `operator->`, `operator[]`, and `typedefs` for compatibility with `std::pointer_traits` and `std::iterator_traits`, the `pmem::obj::persistent_ptr` also has defined methods for persisting its contents. The `pmem::obj::persistent_ptr` can be used in standard library algorithms and containers.

Transactions

Being able to modify more than 8 bytes of storage at a time atomically is imperative for most nontrivial algorithms one might want to use in persistent memory. Commonly, a single logical operation requires multiple stores. For example, an insert into a simple list-based queue requires two separate stores: a tail pointer and the next pointer of the last element. To enable developers to modify larger amounts of data atomically, with respect to power-fail interruptions, the PMDK library provides transaction support in some of its libraries. The C++ language bindings wrap these transactions into two concepts: one, based on the resource acquisition is initialization (RAII) idiom and the other based on a callable `std::function` object. Additionally, because of some C++ standard issues, the scoped transactions come in two flavors: manual and automatic. In this chapter we only describe the approach with `std::function` object. For information about RAII-based transactions, refer to `libpmemobj-cpp` documentation (`https://pmem.io/pmdk/cpp_obj/`).

The method which uses `std::function` is declared as

```
void pmem::obj::transaction::run(pool_base &pop,
    std::function<void ()> tx, Locks&... locks)
```

The locks parameter is a variadic template. Thanks to the std::function, a myriad of types can be passed in to run. One of the preferred ways is to pass a lambda function as the tx parameter. This makes the code compact and easier to analyze. Listing 8-1 shows how lambda can be used to perform work in a transaction.

Listing 8-1. Function object transaction

```
45        // execute a transaction
46        pmem::obj::transaction::run(pop, [&]() {
47            // do transactional work
48        });
```

Of course, this API is not limited to just lambda functions. Any callable target can be passed as tx, such as functions, bind expressions, function objects, and pointers to member functions. Since run is a normal static member function, it has the benefit of being able to throw exceptions. If an exception is thrown during the execution of a transaction, it is automatically aborted, and the active exception is rethrown so information about the interruption is not lost. If the underlying C library fails for any reason, the transaction is also aborted, and a C++ library exception is thrown. The developer is no longer burdened with the task of checking the status of the previous transaction.

libpmemobj-cpp transactions provide an entry point for persistent memory resident synchronization primitives such as pmem::obj::mutex, pmem::obj::shared_mutex and pmem::obj::timed_mutex. libpmemobj ensures that all locks are properly reinitialized when one attempts to acquire a lock for the first time. The use of pmem locks is completely optional, and transactions can be executed without them. The number of supplied locks is arbitrary, and the types can be freely mixed. The locks are held until the end of the given transaction, or the outermost transaction in the case of nesting. This means when transactions are enclosed by a try-catch statement, the locks are released before reaching the catch clause. This is extremely important in case some kind of transaction abort cleanup needs to modify the shared state. In such a case, the necessary locks need to be reacquired in the correct order.

Snapshotting

The C library requires manual snapshots before modifying data in a transaction. The C++ bindings do all of the snapshotting automatically, to reduce the probability of programmer error. The pmem::obj::p template wrapper class is the basic building block for this mechanism. It is designed to work with basic types and not compound types such as classes or PODs (*Plain Old Data*, structures with fields only and without any object-oriented features). This is because it does not define operator->() and there is no possibility to implement operator.(). The implementation of pmem::obj::p is based on the operator=(). Each time the assignment operator is called, the value wrapped by p will be changed, and the library needs to snapshot the old value. In addition to snapshotting, the p<> template ensures the variable is persisted correctly, flushing data if necessary. Listing 8-2 provides an example of using the p<> template.

Listing 8-2. Using the p<> template to persist values correctly

```
39    struct bad_example {
40        int some_int;
41        float some_float;
42    };
43
44    struct good_example {
45        pmem::obj::p<int> pint;
46        pmem::obj::p<float> pfloat;
47    };
48
49    struct root {
50        bad_example bad;
51        good_example good;
52    };
53
54    int main(int argc, char *argv[]) {
55        auto pop = pmem::obj::pool<root>::open("/daxfs/file", "p");
56
57        auto r = pop.root();
58
```

```
59        pmem::obj::transaction::run(pop, [&]() {
60            r->bad.some_int = 10;
61            r->good.pint = 10;
62
63            r->good.pint += 1;
64        });
65
66        return 0;
67    }
```

- Lines 39-42: Here, we declare a bad_example structure with two variables – some_int and some_float. Storing this structure on persistent memory and modifying it are dangerous because data is not snapshotted automatically.

- Lines 44-47: We declare the good_example structure with two p<> type variables – pint and pfloat. This structure can be safely stored on persistent memory as every modification of pint or pfloat in a transaction will perform a snapshot.

- Lines 55-57: Here, we open a persistent memory pool, created already using the pmempool command, and obtain a pointer to the root object stored within the root variable.

- Line 60: We modify the integer value from the bad_example structure. This modification is not safe because we do not add this variable to the transaction; hence it will not be correctly made persistent if there is an unexpected application or system crash or power failure.

- Line 61: Here, we modify integer value wrapped by p<> template. This is safe because operator=() will automatically snapshot the element.

- Line 63: Using arithmetic operators on p<> (if the underlying type supports it) is also safe.

Allocating

As with std::shared_ptr, the pmem::obj::persistent_ptr comes with a set of allocating and deallocating functions. This helps allocate memory and create objects, as well as destroy and deallocate the memory. This is especially important in the case of persistent memory because all allocations and object construction/destruction must be done atomically with respect to power-fail interruptions. The transactional allocations use perfect forwarding and variadic templates for object construction. This makes object creation similar to calling the constructor and identical to std::make_shared. The transactional array creation, however, requires the objects to be default constructible. The created arrays can be multidimensional. The pmem::obj::make_persistent and pmem::obj::make_persistent_array must be called within a transaction; otherwise, an exception is thrown. During object construction, other transactional allocations can be made, and that is what makes this API very flexible. The specifics of persistent memory required the introduction of the pmem::obj::delete_persistent function, which destroys objects and arrays of objects. Since the pmem::obj::persistent_ptr does not automatically handle the lifetime of pointed to objects, the user is responsible for disposing of the ones that are no longer in use. Listing 8-3 shows example of transaction allocation.

Atomic allocations behave differently as they do not return a pointer. Developers must provide a reference to one as the function's argument. Because atomic allocations are not executed in the context of a transaction, the actual pointer assignment must be done through other means. For example, by redo logging the operation. Listing 8-3 also provides an example of atomic allocation.

Listing 8-3. Example of transactional and atomic allocations

```
39    struct my_data {
40        my_data(int a, int b): a(a), b(b) {
41
42        }
43
44        int a;
45        int b;
46    };
47
48    struct root {
49        pmem::obj::persistent_ptr<my_data> mdata;
50    };
51
52    int main(int argc, char *argv[]) {
53        auto pop = pmem::obj::pool<root>::open("/daxfs/file", "tx");
```

```
54
55        auto r = pop.root();
56
57        pmem::obj::transaction::run(pop, [&]() {
58            r->mdata = pmem::obj::make_persistent<my_data>(1, 2);
59        });
60
61        pmem::obj::transaction::run(pop, [&]() {
62            pmem::obj::delete_persistent<my_data>(r->mdata);
63        });
64        pmem::obj::make_persistent_atomic<my_data>(pop, r->mdata,
          2, 3);
65
66        return 0;
67    }
```

- Line 58: Here, we allocate my_data object transactionally. Parameters passed to make_persistent will be forwarded to my_data constructor. Note that assignment to r->mdata will perform a snapshot of old persistent pointer's value.

- Line 62: Here, we delete the my_data object. delete_persistent will call the object's destructor and free the memory.

- Line 64: We allocate my_data object atomically. Calling this function **cannot** be done inside of a transaction.

C++ Standard limitations

The C++ language restrictions and persistent memory programming paradigm imply serious restrictions on objects which may be stored on persistent memory. Applications can access persistent memory with memory-mapped files to take advantage of its byte addressability thanks to libpmemobj and SNIA programming model. No serialization takes place here, so applications must be able to read and modify directly from the persistent memory media even after the application was closed and reopened or after a power failure event.

What does the preceding mean from a C++ and libpmemobj's perspective? There are four major problems:

1. Object lifetime

2. Snapshotting objects in transactions

3. Fixed on-media layout of stored objects

4. Pointers as object members

These four problems will be described in next four sections.

An Object's Lifetime

The lifetime of an object is described in the [basic.life] section of the C++ standard (`https://isocpp.org/std/the-standard`):

> *The lifetime of an object or reference is a runtime property of the object or reference. A variable is said to have vacuous initialization if it is default-initialized and, if it is of class type or a (possibly multi-dimensional) array thereof, that class type has a trivial default constructor. The lifetime of an object of type T begins when:*
>
> *(1.1) storage with the proper alignment and size for type T is obtained, and*
>
> *(1.2) its initialization (if any) is complete (including vacuous initialization) ([dcl.init]), except that if the object is a union member or subobject thereof, its lifetime only begins if that union member is the initialized member in the union ([dcl.init.aggr], [class.base.init]), or as described in [class.union]. The lifetime of an object of type T ends when:*
>
> *(1.3) if T is a non-class type, the object is destroyed, or*
>
> *(1.4) if T is a class type, the destructor call starts, or*
>
> *(1.5) the storage which the object occupies is released, or is reused by an object that is not nested within o ([intro.object]).*

The standard states that properties ascribed to objects apply for a given object only during its lifetime. In this context, the persistent memory programming problem is similar to transmitting data over a network, where the C++ application is given an array of bytes but might be able to recognize the type of object sent. However, the object was not constructed in this application, so using it would result in undefined behavior.

This problem is well known and is being addressed by the WG21 C++ Standards Committee Working Group (https://isocpp.org/std/the-committee and http://www.open-std.org/jtc1/sc22/wg21/).

Currently, there is no possible way to overcome the object-lifetime obstacle and stop relying on undefined behavior from C++ standard's point of view. libpmemobj-cpp is tested and validated with various C++11 compliant compilers and use case scenarios. The only recommendation for libpmemobj-cpp users is that they must keep this limitation in mind when developing persistent memory applications.

Trivial Types

Transactions are the heart of libpmemobj. That is why libpmemobj-cpp was implemented with utmost care while designing the C++ versions so they are as easy to use as possible. Developers do not have to know the implementation details and do not have to worry about snapshotting modified data to make undo log–based transaction works. A special semi-transparent template property class has been implemented to automatically add variable modifications to the transaction undo log, which is described in the "Snapshotting" section.

But what does snapshotting data mean? The answer is very simple, but the consequences for C++ are not. libpmemobj implements snapshotting by copying data of given length from a specified address to another address using memcpy(). If a transaction aborts or a system power loss occurs, the data will be written from the undo log when the memory pool is reopened. Consider a definition of the following C++ object, presented in Listing 8-4, and think about the consequences that a memcpy() has on it.

Listing 8-4. An example showing an unsafe memcpy() on an object

```
35    class nonTriviallyCopyable {
36    private:
37        int* i;
38    public:
39        nonTriviallyCopyable (const nonTriviallyCopyable & from)
40        {
41            /* perform non-trivial copying routine */
42            i = new int(*from.i);
43        }
44    };
```

Deep and shallow copying is the simplest example. The gist of the problem is that by copying the data manually, we may break the inherent behavior of the object which may rely on the copy constructor. Any shared or unique pointer would be another great example – by simple copying it with memcpy(), we break the "deal" we made with that class when we used it, and it may lead to leaks or crashes.

The application must handle many more sophisticated details when it manually copies the contents of an object. The C++11 standard provides a `<type_traits>` type trait and `std::is_trivially_copyable,` which ensure a given type satisfies the requirements of *TriviallyCopyable*. Referring to C++ standard, an object satisfies the *TriviallyCopyable* requirements when

A trivially copyable class is a class that:

— has no non-trivial copy constructors (12.8),

— has no non-trivial move constructors (12.8),

— has no non-trivial copy assignment operators (13.5.3, 12.8),

— has no non-trivial move assignment operators (13.5.3, 12.8), and

— has a trivial destructor (12.4).

A trivial class is a class that has a trivial default constructor (12.1) and is trivially copyable.

[Note: In particular, a trivially copyable or trivial class does not have virtual functions or virtual base classes.]

The C++ standard defines nontrivial methods as follows:

A copy/move constructor for class X is trivial if it is not user-provided and if

— class X has no virtual functions (10.3) and no virtual base classes (10.1), and

— the constructor selected to copy/move each direct base class subobject is trivial, and

— for each non-static data member of X that is of class type (or array thereof), the constructor selected to copy/move that member is trivial;

otherwise, the copy/move constructor is non-trivial.

This means that a copy or move constructor is trivial if it is not user provided. The class has nothing virtual in it, and this property holds recursively for all the members of the class and for the base class. As you can see, the C++ standard and libpmemobj transaction implementation limit the possible objects type to store on persistent memory to satisfy requirements of trivial types, but the layout of our objects must be taken into account.

Object Layout

Object representation, also referred to as the *layout*, might differ between compilers, compiler flags, and application binary interface (ABI). The compiler may do some layout-related optimizations and is free to shuffle order of members with same specifier type – for example, public then protected, then public again. Another problem related to unknown object layout is connected to polymorphic types. Currently there is no reliable and portable way to implement vtable rebuilding after reopening the memory pool, so polymorphic objects cannot be supported with persistent memory.

If we want to store objects on persistent memory using memory-mapped files and to follow the SNIA NVM programming model, we must ensure that the following casting will be always valid:

```
someType A = *reinterpret_cast<someType*>(mmap(...));
```

The bit representation of a stored object type must be always the same, and our application should be able to retrieve the stored object from the memory-mapped file without serialization.

It is possible to ensure that specific types satisfy the aforementioned requirements. C++11 provides another type trait called std::is_standard_layout. The standard mentions that it is useful for communicating with other languages, such as for creating language bindings to native C++ libraries as an example, and that's why a standard-layout class has the same memory layout of the equivalent C struct or union. A general rule is that standard-layout classes must have all non-static data members with the same access control. We mentioned this at the beginning of this section – that a C++ compliant compiler is free to shuffle access ranges of the same class definition.

When using inheritance, only one class in the whole inheritance tree can have non-static data members, and the first non-static data member cannot be of a base class type because this could break aliasing rules. Otherwise, it is not a standard-layout class.

The C++11 standard defines std::is_standard_layout as follows:

A standard-layout class is a class that:

— has no non-static data members of type non-standard-layout class (or array of such types) or reference,

— has no virtual functions (10.3) and no virtual base classes (10.1),

— has the same access control (Clause 11) for all non-static data members,

— has no non-standard-layout base classes,

— either has no non-static data members in the most derived class and at most one base class with non-static data members, or has no base classes with non-static data members, and

— has no base classes of the same type as the first non-static data member.

A standard-layout struct is a standard-layout class defined with the class-key struct or the class-key class.

A standard-layout union is a standard-layout class defined with the class-key union.

[Note: Standard-layout classes are useful for communicating with code written in other programming languages. Their layout is specified in 9.2.]

Having discussed object layouts, we look at another interesting problem with pointer types and how to store them on persistent memory.

Pointers

In previous sections, we quoted parts of the C++ standard. We were describing the limits of types which were safe to snapshot and copy and which we can binary-cast without thinking of fixed layout. But what about pointers? How do we deal with them in our objects as we come to grips with the persistent memory programming model? Consider the code snippet presented in Listing 8-5 which provides an example of a class that uses a volatile pointer as a class member.

Listing 8-5. Example of class with a volatile pointer as a class member

```
39    struct root {
40        int* vptr1;
41        int* vptr2;
42    };
43
44    int main(int argc, char *argv[]) {
45        auto pop = pmem::obj::pool<root>::open("/daxfs/file", "tx");
46
47        auto r = pop.root();
48
49        int a1 = 1;
50
51        pmem::obj::transaction::run(pop, [&](){
52            auto ptr = pmem::obj::make_persistent<int>(0);
53            r->vptr1 = ptr.get();
54            r->vptr2 = &a1;
55        });
56
57        return 0;
58    }
```

- Lines 39-42: We create a `root` structure with two volatile pointers as members.

- Lines 51-52: Our application is assigning, transactionally, two virtual addresses. One to an integer residing on the stack and the second to an integer residing on persistent memory. What will happen if the application crashes or exits after execution of the transaction and we execute the application again? Since the variable `a1` was residing on the stack, the old value vanished. But what is the value assigned to `vptr1`? Even if it resides on persistent memory, the volatile pointer is no longer valid. With ASLR we are not guaranteed to get the same virtual address again if we call `mmap()`. The pointer could point to something, nothing, or garbage.

As shown in the preceding example, it is very important to realize that storing volatile memory pointers in persistent memory is almost always a design error. However, using the pmem::obj::persistent_ptr<> class template is safe. It provides the only way to safely access specific memory after an application crash. However, the pmem::obj::persistent_ptr<> type does not satisfy TriviallyCopyable requirements because of explicitly defined constructors. As a result, an object with a pmem::obj::persistent_ptr<> member will not pass the std::is_trivially_copyable verification check. Every persistent memory developer should always check whether pmem::obj::persistent_ptr<> could be copied in that specific case and that it will not cause errors and persistent memory leaks. Developers should realize that std::is_trivially_copyable is a syntax check only and it does not test the semantics. Using pmem::obj::persistent_ptr<> in this context leads to undefined behavior. There is no single solution to the problem. At the time of writing this book, the C++ standard does not yet fully support persistent memory programming, so developers must ensure that copying pmem::obj::persistent_ptr<> is safe to use in each case.

Limitations Summary

C++11 provides several very useful type traits for persistent memory programming. These are

- template <typename T>

  struct std::is_pod;

- template <typename T>

  struct std::is_trivial;

- template <typename T>

  struct std::is_trivially_copyable;

- template <typename T>

  struct std::is_standard_layout;

They are correlated with each other. The most general and restrictive is the definition of a POD type shown in Figure 8-1.

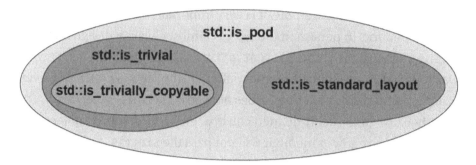

Figure 8-1. *Correlation between persistent memory–related C++ type traits*

We mentioned previously that a persistent memory resident class must satisfy the following requirements:

- `std::is_trivially_copyable`
- `std::is_standard_layout`

Persistent memory developers are free to use more restrictive type traits if required. If we want to use persistent pointers, however, we cannot rely on type traits, and we must be aware of all problems related to copying objects with `memcpy()` and the layout representation of objects. For persistent memory programming, a format description or standardization of the aforementioned concepts and features needs to take place within the C++ standards body group such that it can be officially designed and implemented. Until then, developers must be aware of the restrictions and limitations to manage undefined object-lifetime behavior.

Persistence Simplified

Consider a simple queue implementation, presented in Listing 8-6, which stores elements in volatile DRAM.

Listing 8-6. An implementation of a volatile queue

```
33    #include <cstdio>
34    #include <cstdlib>
35    #include <iostream>
36    #include <string>
37
```

```
38    struct queue_node {
39        int value;
40        struct queue_node *next;
41    };
42
43    struct queue {
44        void
45        push(int value)
46        {
47            auto node = new queue_node;
48            node->value = value;
49            node->next = nullptr;
50
51            if (head == nullptr) {
52                head = tail = node;
53            } else {
54                tail->next = node;
55                tail = node;
56            }
57        }
58
59        int
60        pop()
61        {
62            if (head == nullptr)
63                throw std::out_of_range("no elements");
64
65            auto head_ptr = head;
66            auto value = head->value;
67
68            head = head->next;
69            delete head_ptr;
70
71            if (head == nullptr)
72                tail = nullptr;
73
```

```
74              return value;
75          }
76
77      void
78      show()
79      {
80          auto node = head;
81          while (node != nullptr) {
82              std::cout << "show: " << node->value << std::endl;
83              node = node->next;
84          }
85
86          std::cout << std::endl;
87      }
88
89  private:
90      queue_node *head = nullptr;
91      queue_node *tail = nullptr;
92  };
```

- Lines 38-40: We declare layout of the queue_node structure. It stores an integer value and a pointer to the next node in the list.

- Lines 44-57: We implement push() method which allocates new node and sets its value.

- Lines 59-75: We implement pop() method which deletes the first element in the queue.

- Lines 77-87: The show() method walks the list and prints the contents of each node to standard out.

The preceding queue implementation stores values of type int in a linked list and provides three basic methods: push(), pop(), and show().

In this section, we will demonstrate how to modify your volatile structure to store elements in persistent memory with libpmemobj-cpp bindings. All the modifier methods should provide atomicity and consistency properties which will be guaranteed by the use of transactions.

Changing a volatile application to start taking advantage of persistent memory should rely on modifying structures and classes with only slight modifications to functions. We will begin by modifying the queue_node structure by changing its layout as shown in Listing 8-7.

Listing 8-7. A persistent queue implementation – modifying the queue_node struct

```
38    #include <libpmemobj++/make_persistent.hpp>
39    #include <libpmemobj++/p.hpp>
40    #include <libpmemobj++/persistent_ptr.hpp>
41    #include <libpmemobj++/pool.hpp>
42    #include <libpmemobj++/transaction.hpp>
43
44    struct queue_node {
45        pmem::obj::p<int> value;
46        pmem::obj::persistent_ptr<queue_node> next;
47    };
48
49    struct queue {
...
100   private:
101       pmem::obj::persistent_ptr<queue_node> head = nullptr;
102       pmem::obj::persistent_ptr<queue_node> tail = nullptr;
103   };
```

As you can see, all the modifications are limited to replace the volatile pointers with pmem:obj::persistent_ptr and to start using the p<> property.

Next, we modify a push() method, shown in Listing 8-8.

Listing 8-8. A persistent queue implementation – a persistent push() method

```
50        void
51        push(pmem::obj::pool_base &pop, int value)
52        {
53            pmem::obj::transaction::run(pop, [&]{
54                auto node = pmem::obj::make_persistent<queue_node>();
55                node->value = value;
```

```
56                    node->next = nullptr;
57
58                    if (head == nullptr) {
59                        head = tail = node;
60                    } else {
61                        tail->next = node;
62                        tail = node;
63                    }
64                });
65            }
```

All the modifiers methods must be aware on which persistent memory pool they should operate on. For a single memory pool, this is trivial, but if the application memory maps files from different file systems, we need to keep track of which pool has what data. We introduce an additional argument of type pmem::obj::pool_base to solve this problem. Inside the method definition, we are wrapping the code with a transaction by using a C++ lambda expression, [&], to guarantee atomicity and consistency of modifications. Instead of allocating a new node on the stack, we call pmem::obj::make_persistent<>() to transactionally allocate it on persistent memory.

Listing 8-9 shows the modification of the pop() method.

Listing 8-9. A persistent queue implementation – a persistent pop() method

```
67      int
68      pop(pmem::obj::pool_base &pop)
69      {
70          int value;
71          pmem::obj::transaction::run(pop, [&]{
72              if (head == nullptr)
73                  throw std::out_of_range("no elements");
74
75              auto head_ptr = head;
76              value = head->value;
77
78              head = head->next;
79              pmem::obj::delete_persistent<queue_node>(head_ptr);
80
```

```
81                    if (head == nullptr)
82                        tail = nullptr;
83            });
84
85            return value;
86        }
```

The logic of pop() is wrapped within a libpmemobj-cpp transaction. The only additional modification is to exchange call to volatile delete with transactional pmem::obj::delete_persistent<>().

The show() method does not modify anything on either volatile DRAM or persistent memory, so we do not need to make any changes to it since the pmem:obj::persistent_ptr implementation provides operator->.

To start using the persistent version of this queue example, our application can associate it with a root object. Listing 8-10 presents an example application that uses our persistent queue.

Listing 8-10. Example of application that uses a persistent queue

```
39    #include "persistent_queue.hpp"
40
41    enum queue_op {
42        PUSH,
43        POP,
44        SHOW,
45        EXIT,
46        MAX_OPS,
47    };
48
49    const char *ops_str[MAX_OPS] = {"push", "pop", "show", "exit"};
50
51    queue_op
52    parse_queue_ops(const std::string &ops)
53    {
54        for (int i = 0; i < MAX_OPS; i++) {
55            if (ops == ops_str[i]) {
56                return (queue_op)i;
```

```
57                 }
58             }
59         return MAX_OPS;
60     }
61
62     int
63     main(int argc, char *argv[])
64     {
65         if (argc < 2) {
66             std::cerr << "Usage: " << argv[0] << " path_to_pool"
                   << std::endl;
67             return 1;
68         }
69
70         auto path = argv[1];
71         pmem::obj::pool<queue> pool;
72
73         try {
74             pool = pmem::obj::pool<queue>::open(path, "queue");
75         } catch(pmem::pool_error &e) {
76             std::cerr << e.what() << std::endl;
77             std::cerr << "To create pool run: pmempool create obj
                   --layout=queue -s 100M path_to_pool" << std::endl;
78         }
79
80         auto q = pool.root();
81
82         while (1) {
83             std::cout << "[push value|pop|show|exit]" << std::endl;
84
85             std::string command;
86             std::cin >> command;
87
88             // parse string
89             auto ops = parse_queue_ops(std::string(command));
90
```

```
 91        switch (ops) {
 92            case PUSH: {
 93                int value;
 94                std::cin >> value;
 95
 96                q->push(pool, value);
 97
 98                break;
 99            }
100            case POP: {
101                std::cout << q->pop(pool) << std::endl;
102                break;
103            }
104            case SHOW: {
105                q->show();
106                break;
107            }
108            case EXIT: {
109                exit(0);
110            }
111            default: {
112                std::cerr << "unknown ops" << std::endl;
113                exit(0);
114            }
115        }
116    }
117 }
```

The Ecosystem

The overall goal for the libpmemobj C++ bindings was to create a friendly and less error-prone API for persistent memory programming. Even with persistent memory pool allocators, a convenient interface for creating and managing transactions, auto-snapshotting class templates and smart persistent pointers, and designing an application with persistent memory usage may still prove challenging without a lot of niceties that the C++ programmers are used to. The natural step forward to make persistent programming easier was to provide programmers with efficient and useful containers.

Persistent Containers

The C++ standard library containers collection is something that persistent memory programmers may want to use. Containers manage the lifetime of held objects through allocation/creation and deallocation/destruction with the use of allocators. Implementing custom persistent allocator for C++ STL (Standard Template Library) containers has two main downsides:

- Implementation details:

    - STL containers do not use algorithms optimal for a persistent memory programming point of view.

    - Persistent memory containers should have durability and consistency properties, while not every STL method guarantees strong exception safety.

    - Persistent memory containers should be designed with an awareness of fragmentation limitations.

- Memory layout:

    - The STL does not guarantee that the container layout will remain unchanged in new library versions.

Due to these obstacles, the `libpmemobj-cpp` contains the set of custom, implemented-from-scratch, containers with optimized on-media layouts and algorithms to fully exploit the potential and features of persistent memory. These methods guarantee atomicity, consistency, and durability. Besides specific internal implementation details, `libpmemobj-cpp` persistent memory containers have a well-known STL-like interface, and they work with STL algorithms.

Examples of Persistent Containers

Since the main goal for the libpmemobj-cpp design is to focus modifications to volatile programs on data structures and not on the code, the use of libpmemobj-cpp persistent containers is almost the same as for their STL counterparts. Listing 8-11 shows a persistent vector example to showcase this.

Listing 8-11. Allocating a vector transactionally using persistent containers

```
33    #include <libpmemobj++/make_persistent.hpp>
34    #include <libpmemobj++/transaction.hpp>
35    #include <libpmemobj++/persistent_ptr.hpp>
36    #include <libpmemobj++/pool.hpp>
37    #include "libpmemobj++/vector.hpp"
38
39    using vector_type = pmem::obj::experimental::vector<int>;
40
41    struct root {
42            pmem::obj::persistent_ptr<vector_type> vec_p;
43    };
44

      ...
63
64    /* creating pmem::obj::vector in transaction */
65    pmem::obj::transaction::run(pool, [&] {
66        root->vec_p = pmem::obj::make_persistent<vector_type>
          (/* optional constructor arguments */);
67    });
68
69    vector_type &pvector = *(root->vec_p);
```

Listing 8-11 shows that a pmem::obj::vector must be created and allocated in persistent memory using transaction to avoid an exception being thrown. The vector type constructor may construct an object by internally opening another transaction. In this case, an inner transaction will be flattened to an outer one. The interface and semantics of pmem::obj::vector are similar to that of std::vector, as Listing 8-12 demonstrates.

Listing 8-12. Using persistent containers

```
71          pvector.reserve(10);
72          assert(pvector.size() == 0);
73          assert(pvector.capacity() == 10);
74
75          pvector = {0, 1, 2, 3, 4};
76          assert(pvector.size() == 5);
77          assert(pvector.capacity() == 10);
78
79          pvector.shrink_to_fit();
80          assert(pvector.size() == 5);
81          assert(pvector.capacity() == 5);
82
83          for (unsigned i = 0; i < pvector.size(); ++i)
84              assert(pvector.const_at(i) == static_cast<int>(i));
85
86          pvector.push_back(5);
87          assert(pvector.const_at(5) == 5);
88          assert(pvector.size() == 6);
89
90          pvector.emplace(pvector.cbegin(), pvector.back());
91          assert(pvector.const_at(0) == 5);
92          for (unsigned i = 1; i < pvector.size(); ++i)
93              assert(pvector.const_at(i) == static_cast<int>(i - 1));
```

Every method that modifies persistent memory containers does so inside an implicit transaction to guarantee full exception safety. If any of these methods are called inside the scope of another transaction, the operation is performed in the context of that transaction; otherwise, it is atomic in its own scope.

Iterating over pmem::obj::vector works exactly the same as std::vector. We can use the range-based indexing operator for loops or iterators. The pmem::obj::vector can also be processed using std::algorithms, as shown in Listing 8-13.

Listing 8-13. Iterating over persistent container and compatibility with STD algorithms

```
95        std::vector<int> stdvector = {5, 4, 3, 2, 1};
96        pvector = stdvector;
97
98        try {
99            pmem::obj::transaction::run(pool, [&] {
100               for (auto &e : pvector)
101                   e++;
102               /* 6, 5, 4, 3, 2 */
103
104               for (auto it = pvector.begin();
                  it != pvector.end(); it++)
105                   *it += 2;
106               /* 8, 7, 6, 5, 4 */
107
108               for (unsigned i = 0; i < pvector.size(); i++)
109                   pvector[i]--;
110               /* 7, 6, 5, 4, 3 */
111
112               std::sort(pvector.begin(), pvector.end());
113               for (unsigned i = 0; i < pvector.size(); ++i)
114                   assert(pvector.const_at(i) == static_cast<int>
                  (i + 3));
115
116               pmem::obj::transaction::abort(0);
117           });
118       } catch (pmem::manual_tx_abort &) {
119           /* expected transaction abort */
120       } catch (std::exception &e) {
121           std::cerr << e.what() << std::endl;
122       }
123
```

```
124        assert(pvector == stdvector); /* pvector element's value was
           rolled back */
125
126        try {
127            pmem::obj::delete_persistent<vector_type>(&pvector);
128        } catch (std::exception &e) {
129        }
```

If an active transaction exists, elements accessed using any of the preceding methods
are snapshotted. When iterators are returned by begin() and end(), snapshotting
happens during the iterator dereferencing phase. Note that snapshotting is done only
for mutable elements. In the case of constant iterators or constant versions of indexing
operator, nothing is added to the transaction. That is why it is essential to use const
qualified function overloads such as cbegin() or cend() whenever possible. If an object
snapshot occurs in the current transaction, a second snapshot of the same memory
address will not be performed and thus will not have performance overhead. This will
reduce the number of snapshots and can significantly reduce the performance impact
of transactions. Note also that pmem::obj::vector does define convenient constructors
and compare operators that take std::vector as an argument.

Summary

This chapter describes the libpmemobj-cpp library. It makes creating applications less
error prone, and its similarity to standard C++ API makes it easier to modify existing
volatile programs to use persistent memory. We also list the limitations of this library
and the problems you must consider during development.

CHAPTER 9

pmemkv: A Persistent In-Memory Key-Value Store

Programming persistent memory is not easy. In several chapters we have described that applications that take advantage of persistent memory must take responsibility for atomicity of operations and consistency of data structures. PMDK libraries like libpmemobj are designed with flexibility and simplicity in mind. Usually, these are conflicting requirements, and one has to be sacrificed for the sake of the other. The truth is that in most cases, an API's flexibility increases its complexity.

In the current cloud computing ecosystem, there is an unpredictable demand for data. Consumers expect web services to provide data with predicable low-latency reliability. Persistent memory's byte addressability and huge capacity characteristics make this technology a perfect fit for the broadly defined cloud environment.

Today, as greater numbers of devices with greater levels of intelligence are connected to various networks, businesses and consumers are finding the cloud to be an increasingly attractive option that enables fast, ubiquitous access to their data. Increasingly, consumers are fine with lower storage capacity on endpoint devices in favor of using the cloud. By 2020, IDC predicts that more bytes will be stored in the public cloud than in consumer devices (Figure 9-1).

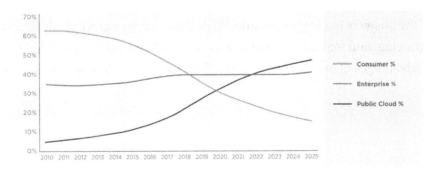

Figure 9-1. *Where is data stored? Source: IDC White Paper – #US44413318*

S. Scargall, *Programming Persistent Memory*, https://doi.org/10.1007/978-1-4842-4932-1_9

The cloud ecosystem, its modularity, and variety of service modes define programming and application deployment as we know it. We call it cloud-native computing, and its popularity results in a growing number of high-level languages, frameworks, and abstraction layers. Figure 9-2 shows the 15 most popular languages on GitHub based on pull requests.

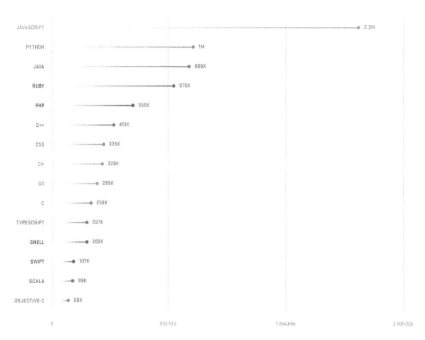

Figure 9-2. *The 15 most popular languages on GitHub by opened pull request (2017). Source:* `https://octoverse.github.com/2017/`

In cloud environments, the platform is typically virtualized, and applications are heavily abstracted as to not make explicit assumptions about low-level hardware details. The question is: how to make programming persistent memory easier in cloud-native environment given the physical devices are local only to a specific server?

One of the answers is a key-value store. This data storage paradigm designed for storing, retrieving, and managing associative arrays with straightforward API can easily utilize the advantages of persistent memory. This is why pmemkv was created.

pmemkv Architecture

There are many key-value data stores available on the market. They have different features and licenses and their APIs are targeting different use cases. However, their core API remains the same. All of them provide methods like put, get, remove, exists, open, and close. At the time we published this book, the most popular key-value data store is Redis. It is available in open source (`https://redis.io/`) and enterprise (`https://redislabs.com`) versions. DB-Engines (`https://db-engines.com`) shows that Redis has a significantly higher rank than any of its competitors in this sector.

Rank	Name	Score
1.	Redis	144,26
2.	Amazon DynamoDB	56,42
3.	Microsoft Azure Cosmos DB	29,08
4.	Memcached	27,07
5.	Hazelcast	8,27
6.	Aerospike	6,59
7.	Ehcache	6,56
8.	Riak KV	6,06
9.	OrientDB	5,69
10.	ArangoDB	4,66
11.	Ignite	4,26
12.	Oracle NoSQL	3,46
13.	InterSystems Caché	3,30
14.	LevelDB	3,29
15.	Oracle Berkeley DB	3,04

Figure 9-3. *DB-Engines ranking of key-value stores (July 2019). Scoring method:* `https://db-engines.com/en/ranking_definition`. *Source:* `https://db-engines.com/en/ranking/key-value+store`

Pmemkv was created as a separate project not only to complement PMDK's set of libraries with cloud-native support but also to provide a key-value API built for persistent memory. One of the main goals for pmemkv developers was to create friendly environment for open source community to develop new engines with the help of PMDK and to integrate it with other programming languages. Pmemkv uses the same BSD 3-Clause permissive license as PMDK. The native API of pmemkv is C and C++. Other programming language bindings are available such as JavaScript, Java, and Ruby. Additional languages can easily be added.

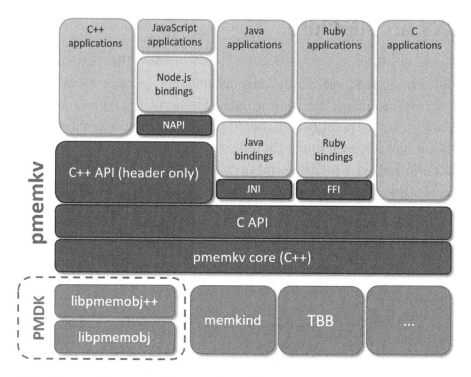

Figure 9-4. *The architecture of pmemkv and programming languages support*

The pmemkv API is similar to most key-value databases. Several storage engines are available for flexibility and functionality. Each engine has different performance characteristics and aims to solve different problems. Because of that, the functionality provided by each engine differs. They can be described by the following characteristics:

- Persistence: Persistent engines guarantee that modifications are retained and power-fail safe, while volatile ones keep its content only for the application lifetime.

- Concurrency: Concurrent engines guarantee that some methods such as get()/put()/remove() are thread-safe.

- Keys' ordering: "Sorted" engines provide range query methods (like get_above()).

What makes pmemkv different from other key-value databases is that it provides direct access to the data. This means reading data from persistent memory does not require a copy into DRAM. This was already mentioned in Chapter 1 and is presented again in Figure 9-5.

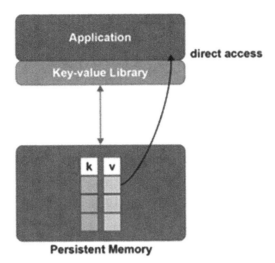

Figure 9-5. *Applications directly accessing data in place using pmemkv*

Having direct access to the data significantly speeds up the application. This benefit is most noticeable in situations where the program is only interested in a part of the data stored in the database. In conventional approaches, this would require copying the whole data in some buffer and returning it to the application. With pmemkv, we provide the application a direct pointer, and the application reads only as much as it is needed.

To make the API fully functional with various engine types, a flexible pmemkv_config structure was introduced. It stores engine configuration options and allows you to tune its behavior. Every engine has documented all supported config parameters. The pmemkv library was designed in a way that engines are pluggable and extendable to support the developers own requirements. Developers are free to modify existing engines or contribute new ones (https://github.com/pmem/pmemkv/blob/master/CONTRIBUTING.md#engines).

Listing 9-1 shows a basic setup of the pmemkv_config structure using the native C API. All the setup code is wrapped around the custom function, config_setup(), which will be used in a phonebook example in the next section. You can see how error handling is solved in pmemkv – all methods, except for pmemkv_close() and pmemkv_errormsg(), return a status. We can obtain error message using the pmemkv_errormsg() function. A complete list of return values can be found in pmemkv man page.

Listing 9-1. pmemkv_config.h – An example of the pmemkv_config structure using the C API

```
1    #include <cstdio>
2    #include <cassert>
3    #include <libpmemkv.h>
4
5    pmemkv_config* config_setup(const char* path, const uint64_t fcreate,
     const uint64_t size) {
6        pmemkv_config *cfg = pmemkv_config_new();
7        assert(cfg != nullptr);
8
9        if (pmemkv_config_put_string(cfg, "path", path) != PMEMKV_STATUS_OK) {
10            fprintf(stderr, "%s", pmemkv_errormsg());
11            return NULL;
12        }
13
14        if (pmemkv_config_put_uint64(cfg, "force_create", fcreate) !=
         PMEMKV_STATUS_OK) {
15            fprintf(stderr, "%s", pmemkv_errormsg());
16            return NULL;
17        }
18
19        if (pmemkv_config_put_uint64(cfg, "size", size) != PMEMKV_STATUS_OK) {
20            fprintf(stderr, "%s", pmemkv_errormsg());
21            return NULL;
22        }
23
24        return cfg;
25    }
```

- Line 5: We define custom function to prepare config and set all required params for engine(s) to use.

- Line 6: We create an instance of C config class. It returns nullptr on failure.

- Line 9-22: All params are put into config (the cfg instance) one after another (using function dedicated for the type), and each is checked if was stored successful (PMEMKV_STATUS_OK is returned when no errors occurred).

A Phonebook Example

Listing 9-2 shows a simple phonebook example implemented using the pmemkv C++ API v0.9. One of the main intentions of pmemkv is to provide a familiar API similar to the other key-value stores. This makes it very intuitive and easy to use. We will reuse the config_setup() function from Listing 9-1.

Listing 9-2. A simple phonebook example using the pmemkv C++ API

```
37    #include <iostream>
38    #include <cassert>
39    #include <libpmemkv.hpp>
40    #include <string>
41    #include "pmemkv_config.h"
42
43    using namespace pmem::kv;
44
45    auto PATH = "/daxfs/kvfile";
46    const uint64_t FORCE_CREATE = 1;
47    const uint64_t SIZE = 1024 * 1024 * 1024; // 1 Gig
48
49    int main() {
50        // Prepare config for pmemkv database
51        pmemkv_config *cfg = config_setup(PATH, FORCE_CREATE, SIZE);
52        assert(cfg != nullptr);
53
54        // Create a key-value store using the "cmap" engine.
55        db kv;
56
57        if (kv.open("cmap", config(cfg)) != status::OK) {
58            std::cerr << db::errormsg() << std::endl;
```

```
59              return 1;
60          }
61
62          // Add 2 entries with name and phone number
63          if (kv.put("John", "123-456-789") != status::OK) {
64              std::cerr << db::errormsg() << std::endl;
65              return 1;
66          }
67          if (kv.put("Kate", "987-654-321") != status::OK) {
68              std::cerr << db::errormsg() << std::endl;
69              return 1;
70          }
71
72          // Count elements
73          size_t cnt;
74          if (kv.count_all(cnt) != status::OK) {
75              std::cerr << db::errormsg() << std::endl;
76              return 1;
77          }
78          assert(cnt == 2);
79
80          // Read key back
81          std::string number;
82          if (kv.get("John", &number) != status::OK) {
83              std::cerr << db::errormsg() << std::endl;
84              return 1;
85          }
86          assert(number == "123-456-789");
87
88          // Iterate through the phonebook
89          if (kv.get_all([](string_view name, string_view number) {
90                  std::cout << "name: " << name.data() <<
91                  ", number: " << number.data() << std::endl;
92                  return 0;
93                  }) != status::OK) {
```

```
94          std::cerr << db::errormsg() << std::endl;
95          return 1;
96      }
97
98      // Remove one record
99      if (kv.remove("John") != status::OK) {
100         std::cerr << db::errormsg() << std::endl;
101         return 1;
102     }
103
104     // Look for removed record
105     assert(kv.exists("John") == status::NOT_FOUND);
106
107     // Try to use one of methods of ordered engines
108     assert(kv.get_above("John", [](string_view key, string_view
        value) {
109         std::cout << "This callback should never be called" <<
            std::endl;
110         return 1;
111     }) == status::NOT_SUPPORTED);
112
113     // Close database (optional)
114     kv.close();
115
116     return 0;
117 }
```

- Line 51: We set the pmemkv_config structure by calling config_
  setup() function introduced in previous section and listing
  (imported with #include "pmemkv_config.h").

- Line 55: Creates a volatile object instance of the class pmem::kv::db
  which provides interface for managing persistent database.

- Line 57: Here, we open the key-value database backed by the *cmap*
  engine using the config parameters. The cmap engine is a persistent
  concurrent hash map engine, implemented in libpmemobj-cpp.

You can read more about *cmap* engine internal algorithms and data structures in Chapter 13.

- Line 58: The `pmem::kv::db` class provides a static `errormsg()` method for extended error messages. In this example, we use the `errormsg()` function as a part of the error-handling routine.

- Line 63 and 67: The `put()` method inserts a key-value pair into the database. This function is guaranteed to be implemented by all engines. In this example, we are inserting two key-value pairs into database and compare returned statuses with `status::OK`. It's a recommended way to check if function succeeded.

- Line 74: The `count_all()` has a single argument of type `size_t`. The method returns the number of elements (phonebook entries) stored in the database by the argument variable (`cnt`).

- Line 82: Here, we use the `get()` method to return the value of the "John" key. The value is copied into the user-provided `number` variable. The `get()` function returns `status::OK` on success or an error on failure. This function is guaranteed to be implemented by all engines.

- Line 86: For this example, the expected value of variable `number` for "John" is "123-456-789". If we do not get this value, an assertion error is thrown.

- Line 89: The `get_all()` method used in this example gives the application direct, read-only access to the data. Both key and value variables are references to data stored in persistent memory. In this example, we simply print the name and the number of every visited pair.

- Line 99: Here, we are removing "John" and his phone number from the database by calling the `remove()` method. It is guaranteed to be implemented by all engines.

- Line 105: After removal of the pair "*John, 123-456-789*", we verify if the pair still exists in database. The API method `exists()` checks the existence of an element with given key. If the element is present, `status::OK` is returned; otherwise `status::NOT_FOUND` is returned.

- Line 108: Not every engine provides implementations of all the available API methods. In this example, we used the *cmap* engine, which is unordered engine type. This is why *cmap* does not support the get_above() function (and similarly: get_below(), get_between(), count_above(), count_below(), count_between()). Calling these functions will return status::NOT_SUPPORTED.

- Line 114: Finally, we are calling the close() method to close database. Calling this function is optional because kv was allocated on the stack and all necessary destructors will be called automatically, just like for the other variables residing on stack.

Bringing Persistent Memory Closer to the Cloud

We will rewrite the phonebook example using the JavaScript language bindings. There are several language bindings available for pmemkv – JavaScript, Java, Ruby, and Python. However, not all provide the same API functionally equivalent to the native C and C++ counterparts. Listing 9-3 shows an implementation of the phonebook application written using JavaScript language bindings API.

Listing 9-3. A simple phonebook example written using the JavaScript bindings for pmemkv v0.8

```
1    const Database = require('./lib/all');
2
3    function assert(condition) {
4        if (!condition) throw new Error('Assert failed');
5    }
6
7    console.log('Create a key-value store using the "cmap" engine');
8    const db = new Database('cmap', '{"path":"/daxfs/
     kvfile","size":1073741824, "force_create":1}');
9
10   console.log('Add 2 entries with name and phone number');
11   db.put('John', '123-456-789');
12   db.put('Kate', '987-654-321');
13
```

```
14     console.log('Count elements');
15     assert(db.count_all == 2);
16
17     console.log('Read key back');
18     assert(db.get('John') === '123-456-789');
19
20     console.log('Iterate through the phonebook');
21     db.get_all((k, v) => console.log(`  name: ${k}, number: ${v}`));
22
23     console.log('Remove one record');
24     db.remove('John');
25
26     console.log('Lookup of removed record');
27     assert(!db.exists('John'));
28
29     console.log('Stopping engine');
30     db.stop();
```

The goal of higher-level pmemkv language bindings is to make programming persistent memory even easier and to provide a convenient tool for developers of cloud software.

Summary

In this chapter, we have shown how a familiar key-value data store is an easy way for the broader cloud software developer audience to use persistent memory and directly access the data in place. The modular design, flexible engine API, and integration with many of the most popular cloud programming languages make pmemkv an intuitive choice for cloud-native software developers. As an open source and lightweight library, it can easily be integrated into existing applications to immediately start taking advantage of persistent memory.

Some of the pmemkv engines are implemented using libpmemobj-cpp that we described in Chapter 8. The implementation of such engines provides real-world examples for developers to understand how to use PMDK (and related libraries) in applications.

CHAPTER 10

Volatile Use of Persistent Memory

Introduction

This chapter discusses how applications that require a large quantity of volatile memory can leverage high-capacity persistent memory as a complementary solution to dynamic random-access memory (DRAM).

Applications that work with large data sets, like in-memory databases, caching systems, and scientific simulations, are often limited by the amount of volatile memory capacity available in the system or the cost of the DRAM required to load a complete data set. Persistent memory provides a high capacity memory tier to solve these memory-hungry application problems.

In the memory-storage hierarchy (described in Chapter 1), data is stored in tiers with frequently accessed data placed in DRAM for low-latency access, and less frequently accessed data is placed in larger capacity, higher latency storage devices. Examples of such solutions include Redis on Flash (`https://redislabs.com/redis-enterprise/technology/redis-on-flash/`) and Extstore for Memcached (`https://memcached.org/blog/extstore-cloud/`).

For memory-hungy applications that do not require persistence, using the larger capacity persistent memory as volatile memory provides new opportunities and solutions.

Using persistent memory as a volatile memory solution is advantageous when an application:

- Has control over data placement between DRAM and other storage tiers within the system

- Does not need to persist data

© The Author(s) 2020
S. Scargall, *Programming Persistent Memory*, https://doi.org/10.1007/978-1-4842-4932-1_10

- Can use the native latencies of persistent memory, which may be slower than DRAM but are faster than non-volatile memory express (NVMe) solid-state drives (SSDs).

Background

Applications manage different kinds of data structures such as user data, key-value stores, metadata, and working buffers. Architecting a solution that uses tiered memory and storage may enhance application performance, for example, placing objects that are accessed frequently and require low-latency access in DRAM while storing objects that require larger allocations that are not as latency-sensitive on persistent memory. Traditional storage devices are used to provide persistence.

Memory Allocation

As described in Chapters 1 through 3, persistent memory is exposed to the application using memory-mapped files on a persistent memory-aware file system that provides direct access to the application. Since `malloc()` and `free()` do not operate on different types of memory or memory-mapped files, an interface is needed that provides `malloc()` and `free()` semantics for multiple memory types. This interface is implemented as the memkind library (`http://memkind.github.io/memkind/`).

How it Works

The memkind library is a user-extensible heap manager built on top of `jemalloc`, which enables partitioning of the heap between multiple *kinds* of memory. Memkind was created to support different kinds of memory when high bandwidth memory (HBM) was introduced. A PMEM *kind* was introduced to support persistent memory.

Different "kinds" of memory are defined by the operating system memory policies that are applied to virtual address ranges. Memory characteristics supported by memkind without user extension include the control of non-uniform memory access (NUMA) and page sizes. Figure 10-1 shows an overview of libmemkind components and hardware support.

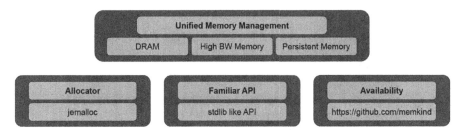

Figure 10-1. *An overview of the memkind components and hardware support*

The memkind library serves as a wrapper that redirects memory allocation requests from an application to an allocator that manages the heap. At the time of publication, only the jemalloc allocator is supported. Future versions may introduce and support multiple allocators. Memkind provides jemalloc with different kinds of memory: A *static kind* is created automatically, whereas a *dynamic kind* is created by an application using memkind_create_kind().

Supported "Kinds" of Memory

The dynamic PMEM kind is best used with memory-addressable persistent storage through a DAX-enabled file system that supports load/store operations that are not paged via the system page cache. For the PMEM kind, the memkind library supports the traditional malloc/free-like interfaces on a memory-mapped file. When an application calls memkind_create_kind() with PMEM, a temporary file (tmpfile(3)) is created on a mounted DAX file system and is memory-mapped into the application's virtual address space. This temporary file is deleted automatically when the program terminates, giving the perception of volatility.

Figure 10-2 shows memory mappings from two memory sources: DRAM (MEMKIND_DEFAULT) and persistent memory (PMEM_KIND).

For allocations from DRAM, rather than using the common malloc(), the application can call memkind_malloc() with the *kind* argument set to MEMKIND_DEFAULT. MEMKIND_DEFAULT is a static kind that uses the operating system's default page size for allocations. Refer to the memkind documentation for large and huge page support.

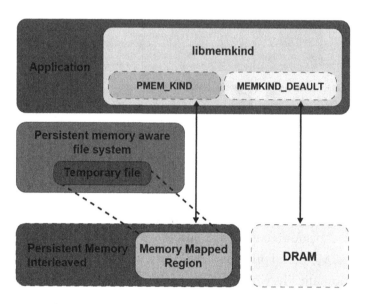

Figure 10-2. *An application using different "kinds" of memory*

When using `libmemkind` with DRAM and persistent memory, the key points to understand are:

- Two pools of memory are available to the application, one from DRAM and another from persistent memory.

- Both pools of memory can be accessed simultaneously by setting the kind type to `PMEM_KIND` to use persistent memory and `MEMKIND_DEFAULT` to use DRAM.

- `jemalloc` is the single memory allocator used to manage all kinds of memory.

- The memkind library is a wrapper around `jemalloc` that provides a unified API for allocations from different kinds of memory.

- `PMEM_KIND` memory allocations are provided by a temporary file (tmpfile(3)) created on a persistent memory-aware file system. The file is destroyed when the application exits. Allocations are not persistent.

- Using `libmemkind` for persistent memory requires simple modifications to the application.

The memkind API

The memkind API functions related to persistent memory programming are shown in Listing 10-1 and described in this section. The complete memkind API is available in the memkind man pages (http://memkind.github.io/memkind/man_pages/memkind.html).

Listing 10-1. Persistent memory-related memkind API functions

KIND CREATION MANAGEMENT:
```
int memkind_create_pmem(const char *dir, size_t max_size, memkind_t *kind);
int memkind_create_pmem_with_config(struct memkind_config *cfg, memkind_t
*kind);
memkind_t memkind_detect_kind(void *ptr);
int memkind_destroy_kind(memkind_t kind);
```

KIND HEAP MANAGEMENT:
```
void *memkind_malloc(memkind_t kind, size_t size);
void *memkind_calloc(memkind_t kind, size_t num, size_t size);
void *memkind_realloc(memkind_t kind, void *ptr, size_t size);
void memkind_free(memkind_t kind, void *ptr);
size_t memkind_malloc_usable_size(memkind_t kind, void *ptr);
memkind_t memkind_detect_kind(void *ptr);
```

KIND CONFIGURATION MANAGEMENT:
```
struct memkind_config *memkind_config_new();
void memkind_config_delete(struct memkind_config *cfg);
void memkind_config_set_path(struct memkind_config *cfg, const char
*pmem_dir);
void memkind_config_set_size(struct memkind_config *cfg, size_t pmem_size);
void memkind_config_set_memory_usage_policy(struct memkind_config *cfg,
memkind_mem_usage_policy policy);
```

Kind Management API

The memkind library supports a plug-in architecture to incorporate new memory kinds, which are referred to as dynamic kinds. The memkind library provides the API to create and manage the heap for the dynamic kinds.

Kind Creation

Use the memkind_create_pmem() function to create a PMEM *kind* of memory from a file-backed source. This file is created as a tmpfile(3) in a specified directory (PMEM_DIR) and is unlinked, so the file name is not listed under the directory. The temporary file is automatically removed when the program terminates.

Use memkind_create_pmem() to create a fixed or dynamic heap size depending on the application requirement. Additionally, configurations can be created and supplied rather than passing in configuration options to the *_create_* function.

Creating a Fixed-Size Heap

Applications that require a fixed amount of memory can specify a nonzero value for the PMEM_MAX_SIZE argument to memkind_create_pmem(), shown below. This defines the size of the memory pool to be created for the specified kind of memory. The value of PMEM_MAX_SIZE should be less than the available capacity of the file system specified in PMEM_DIR to avoid ENOMEM or ENOSPC errors. An internal data structure struct memkind is populated internally by the library and used by the memory management functions.

```
int memkind_create_pmem(PMEM_DIR, PMEM_MAX_SIZE, &pmem_kind)
```

The arguments to memkind_create_pmem() are

- PMEM_DIR is the directory where the temp file is created.

- PMEM_MAX_SIZE is the size, in bytes, of the memory region to be passed to jemalloc.

- &pmem_kind is the address of a memkind data structure.

If successful, memkind_create_pmem() returns zero. On failure, an error number is returned that memkind_error_message() can convert to an error message string. Listing 10-2 shows how a 32MiB PMEM kind is created on a /daxfs file system. Included in this listing is the definition of memkind_fatal() to print a memkind error message and exit. The rest of the examples in this chapter assume this routine is defined as shown below.

Listing 10-2. Creating a 32MiB PMEM kind

```
void memkind_fatal(int err)
{
    char error_message[MEMKIND_ERROR_MESSAGE_SIZE];
```

```
    memkind_error_message(err, error_message,
        MEMKIND_ERROR_MESSAGE_SIZE);
    fprintf(stderr, "%s\n", error_message);
    exit(1);
}

/* ... in main() ... */

#define PMEM_MAX_SIZE (1024 * 1024 * 32)

struct memkind *pmem_kind;
int err;

// Create PMEM memory pool with specific size
err = memkind_create_pmem("/daxfs",PMEM_MAX_SIZE, &pmem_kind);
if (err) {
    memkind_fatal(err);
}
```

You can also create a heap with a specific configuration using the function memkind_
create_pmem_with_config(). This function uses a memkind_config structure with
optional parameters such as size, file path, and memory usage policy. Listing 10-3
shows how to build a test_cfg using memkind_config_new(), then passing that
configuration to memkind_create_pmem_with_config() to create a PMEM kind. We use
the same path and size parameters from the Listing 10-2 example for comparison.

Listing 10-3. Creating PMEM kind with configuration

```
struct memkind_config *test_cfg = memkind_config_new();
memkind_config_set_path(test_cfg, "/daxfs");
memkind_config_set_size(test_cfg, 1024 * 1024 * 32);
memkind_config_set_memory_usage_policy(test_cfg, MEMKIND_MEM_USAGE_POLICY_
CONSERVATIVE);

// create a PMEM partition with specific configuration
err = memkind_create_pmem_with_config(test_cfg, &pmem_kind);
if (err) {
    memkind_fatal(err);
}
```

Creating a Variable Size Heap

When PMEM_MAX_SIZE is set to zero, as shown below, allocations are satisfied as long as the temporary file can grow. The maximum heap size growth is limited by the capacity of the file system mounted under the PMEM_DIR argument.

memkind_create_pmem(PMEM_DIR, 0, &pmem_kind)

The arguments to memkind_create_pmem() are:

- PMEM_DIR is the directory where the temp file is created.
- PMEM_MAX_SIZE is 0.
- &pmem_kind is the address of a memkind data structure.

If the PMEM kind is created successfully, memkind_create_pmem() returns zero. On failure, memkind_error_message() can be used to convert an error number returned by memkind_create_pmem() to an error message string, as shown in the memkind_fatal() routine in Listing 10-2.

Listing 10-4 shows how to create a PMEM kind with variable size.

Listing 10-4. Creating a PMEM kind with variable size

```
struct memkind *pmem_kind;
int err;
err = memkind_create_pmem("/daxfs",0,&pmem_kind);
if (err) {
    memkind_fatal(err);
}
```

Detecting the Memory Kind

Memkind supports both automatic detection of the kind as well as a function to detect the kind associated with a memory referenced by a pointer.

Automatic Kind Detection

Automatically detecting the kind of memory is supported to simplify code changes when using libmemkind. Thus, the memkind library will automatically retrieve the *kind* of memory pool the allocation was made from, so the heap management functions listed in Table 10-1 can be called without specifying the kind.

Table 10-1. *Automatic kind detection functions and their equivalent specified kind functions and operations*

Operation	Memkind API with Kind	Memkind API Using Automatic Detection
free	memkind_free(kind, ptr)	memkind_free(NULL, ptr)
realloc	memkind_realloc(kind, ptr, size)	memkind_realloc(NULL, ptr, size)
Get size of allocated memory	memkind_malloc_usable_size(kind, ptr)	memkind_malloc_usable_size(NULL, ptr)

The memkind library internally tracks the kind of a given object from the allocator metadata. However, to get this information, some of the operations may need to acquire a lock to prevent accesses from other threads, which may negatively affect the performance in a multithreaded environment.

Memory Kind Detection

Memkind also provides the memkind_detect_kind() function, shown below, to query and return the kind of memory referenced by the pointer passed into the function. If the input pointer argument is NULL, the function returns NULL. The input pointer argument passed into memkind_detect_kind() must have been returned by a previous call to memkind_malloc(), memkind_calloc(), memkind_realloc(), or memkind_posix_memalign().

memkind_t memkind_detect_kind(void *ptr)

Similar to the automatic detection approach, this function has nontrivial performance overhead. Listing 10-5 shows how to detect the kind type.

Listing 10-5. pmem_detect_kind.c – how to automatically detect the 'kind' type

```
73  err = memkind_create_pmem(path, 0, &pmem_kind);
74  if (err) {
75      memkind_fatal(err);
76  }
77
```

```
78  /* do some allocations... */
79  buf0 = memkind_malloc(pmem_kind, 1000);
80  buf1 = memkind_malloc(MEMKIND_DEFAULT, 1000);
81
82  /* look up the kind of an allocation */
83  if (memkind_detect_kind(buf0) == MEMKIND_DEFAULT) {
84      printf("buf0 is DRAM\n");
85  } else {
86      printf("buf0 is pmem\n");
87  }
```

Destroying Kind Objects

Use the memkind_destroy_kind() function, shown below, to delete the kind object that was previously created using the memkind_create_pmem() or memkind_create_pmem_with_config() function.

int memkind_destroy_kind(memkind_t kind);

Using the same pmem_detect_kind.c code from Listing 10-5, Listing 10-6 shows how the kind is destroyed before the program exits.

Listing 10-6. Destroying a kind object

```
89      err = memkind_destroy_kind(pmem_kind);
90      if (err) {
91          memkind_fatal(err);
92      }
```

When the kind returned by memkind_create_pmem() or memkind_create_pmem_with_config() is successfully destroyed, all the allocated memory for the kind object is freed.

Heap Management API

The heap management functions described in this section have an interface modeled on the ISO C standard API, with an additional "kind" parameter to specify the memory type used for allocation.

Allocating Memory

The memkind library provides memkind_malloc(), memkind_calloc(), and memkind_realloc() functions for allocating memory, defined as follows:

```
void *memkind_malloc(memkind_t kind, size_t size);
void *memkind_calloc(memkind_t kind, size_t num, size_t size);
void *memkind_realloc(memkind_t kind, void *ptr, size_t size);
```

memkind_malloc() allocates size bytes of uninitialized memory of the specified kind. The allocated space is suitably aligned (after possible pointer coercion) for storage of any object type. If size is 0, then memkind_malloc() returns NULL.

memkind_calloc() allocates space for num objects, each is size bytes in length. The result is identical to calling memkind_malloc() with an argument of num * size. The exception is that the allocated memory is explicitly initialized to zero bytes. If num or size is 0, then memkind_calloc() returns NULL.

memkind_realloc() changes the size of the previously allocated memory referenced by ptr to size bytes of the specified kind. The contents of the memory remain unchanged, up to the lesser of the new and old sizes. If the new size is larger, the contents of the newly allocated portion of the memory are undefined. If successful, the memory referenced by ptr is freed, and a pointer to the newly allocated memory is returned.

The code example in Listing 10-7 shows how to allocate memory from DRAM and persistent memory (pmem_kind) using memkind_malloc(). Rather than using the common C library malloc() for DRAM and memkind_malloc() for persistent memory, we recommend using a single library to simplify the code.

Listing 10-7. An example of allocating memory from both DRAM and persistent memory

```
/*
 * Allocates 100 bytes using appropriate "kind"
 * of volatile memory
 */
```

```
// Create a PMEM memory pool with a specific size
  err = memkind_create_pmem(path, PMEM_MAX_SIZE, &pmem_kind);
  if (err) {
      memkind_fatal(err);
  }
  char *pstring = memkind_malloc(pmem_kind, 100);
  char *dstring = memkind_malloc(MEMKIND_DEFAULT, 100);
```

Freeing Allocated Memory

To avoid memory leaks, allocated memory can be freed using the memkind_free()
function, defined as:

```
void memkind_free(memkind_t kind, void *ptr);
```

   memkind_free() causes the allocated memory referenced by ptr to be made
available for future allocations. This pointer must be returned by a previous call to
memkind_malloc(), memkind_calloc(), memkind_realloc(), or memkind_posix_
memalign(). Otherwise, if memkind_free(kind, ptr) was previously called, undefined
behavior occurs. If ptr is NULL, no operation is performed. In cases where the kind is
unknown in the context of the call to memkind_free(), NULL can be given as the kind
specified to memkind_free(), but this will require an internal lookup for the correct kind.
Always specify the correct kind because the lookup for kind could result in a serious
performance penalty.

   Listing 10-8 shows four examples of memkind_free() being used. The first two specify
the kind, and the second two use NULL to detect the kind automatically.

Listing 10-8. Examples of memkind_free() usage

```
/* Free the memory by specifying the kind */
memkind_free(MEMKIND_DEFAULT, dstring);
memkind_free(PMEM_KIND, pstring);

/* Free the memory using automatic kind detection */
memkind_free(NULL, dstring);
memkind_free(NULL, pstring);
```

Kind Configuration Management

You can also create a heap with a specific configuration using the function `memkind_create_pmem_with_config()`. This function requires completing a `memkind_config` structure with optional parameters such as size, path to file, and memory usage policy.

Memory Usage Policy

In jemalloc, a runtime option called `dirty_decay_ms` determines how fast it returns unused memory back to the operating system. A shorter decay time purges unused memory pages faster, but the purging costs CPU cycles. Trade-offs between memory and CPU cycles needed for this operation should be carefully thought out before using this parameter.

The memkind library supports two policies related to this feature:

1. MEMKIND_MEM_USAGE_POLICY_DEFAULT

2. MEMKIND_MEM_USAGE_POLICY_CONSERVATIVE

The minimum and maximum values for `dirty_decay_ms` using the `MEMKIND_MEM_USAGE_POLICY_DEFAULT` are 0ms to 10,000ms for arenas assigned to a PMEM kind. Setting `MEMKIND_MEM_USAGE_POLICY_CONSERVATIVE` sets shorter decay times to purge unused memory faster, reducing memory usage. To define the memory usage policy, use `memkind_config_set_memory_usage_policy()`, shown below:

```
void memkind_config_set_memory_usage_policy (struct memkind_config *cfg,
memkind_mem_usage_policy policy );
```

- `MEMKIND_MEM_USAGE_POLICY_DEFAULT` is the default memory usage policy.

- `MEMKIND_MEM_USAGE_POLICY_CONSERVATIVE` allows changing the `dirty_decay_ms` parameter.

Listing 10-9 shows how to use `memkind_config_set_memory_usage_policy()` with a custom configuration.

Listing 10-9. An example of a custom configuration and memory policy use

```
73  struct memkind_config *test_cfg =
74      memkind_config_new();
75  if (test_cfg == NULL) {
76      fprintf(stderr,
77          "memkind_config_new: out of memory\n");
78      exit(1);
79  }
80
81  memkind_config_set_path(test_cfg, path);
82  memkind_config_set_size(test_cfg, PMEM_MAX_SIZE);
83  memkind_config_set_memory_usage_policy(test_cfg,
84      MEMKIND_MEM_USAGE_POLICY_CONSERVATIVE);
85
86  // Create PMEM partition with the configuration
87  err = memkind_create_pmem_with_config(test_cfg,
88      &pmem_kind);
89  if (err) {
90      memkind_fatal(err);
91  }
```

Additional memkind Code Examples

The memkind source tree contains many additional code examples, available on GitHub at https://github.com/memkind/memkind/tree/master/examples.

C++ Allocator for PMEM Kind

A new pmem::allocator class template is created to support allocations from persistent memory, which conforms to C++11 allocator requirements. It can be used with C++ compliant data structures from:

- Standard Template Library (STL)

- Intel® Threading Building Blocks (Intel® TBB) library

The pmem::allocator class template uses the memkind_create_pmem() function described previously. This allocator is stateful and has no default constructor.

pmem::allocator methods

```
pmem::allocator(const char *dir, size_t max_size);
pmem::allocator(const std::string& dir, size_t max_size) ;
template <typename U> pmem::allocator<T>::allocator(const
pmem::allocator<U>&);
template <typename U> pmem::allocator(allocator<U>&& other);
pmem::allocator<T>::~allocator();
T* pmem::allocator<T>::allocate(std::size_t n) const;
void pmem::allocator<T>::deallocate(T* p, std::size_t n) const ;
template <class U, class... Args> void pmem::allocator<T>::construct(U* p,
Args... args) const;
void pmem::allocator<T>::destroy(T* p) const;
```

For more information about the pmem::allocator class template, refer to the pmem allocator(3) man page.

Nested Containers

Multilevel containers such as a vector of lists, tuples, maps, strings, and so on pose challenges in handling the nested objects.

Imagine you need to create a vector of strings and store it in persistent memory. The challenges – and their solutions – for this task include:

1. Challenge: The std::string cannot be used for this purpose because it is an alias of the std::basic_string. The std::allocator requires a new alias that uses pmem:allocator.

   Solution: A new alias called pmem_string is defined as a typedef of std::basic_string when created with pmem::allocator.

2. Challenge: How to ensure that an outermost vector will properly construct nested pmem_string with a proper instance of pmem::allocator.

    Solution: From C++11 and later, the std::scoped_allocator_adaptor class template can be used with multilevel containers. The purpose of this adaptor is to correctly initialize stateful allocators in nested containers, such as when all levels of a nested container must be placed in the same memory segment.

C++ Examples

This section presents several full-code examples demonstrating the use of libmemkind using C and C++.

Using the pmem::allocator

As mentioned earlier, you can use pmem::allocator with any STL-like data structure. The code sample in Listing 10-10 includes a pmem_allocator.h header file to use pmem::allocator.

Listing 10-10. pmem_allocator.cpp: using pmem::allocator with std:vector

```
37  #include <pmem_allocator.h>
38  #include <vector>
39  #include <cassert>
40
41  int main(int argc, char *argv[]) {
42      const size_t pmem_max_size = 64 * 1024 * 1024; //64 MB
43      const std::string pmem_dir("/daxfs");
44
45      // Create allocator object
46      libmemkind::pmem::allocator<int>
47          alc(pmem_dir, pmem_max_size);
48
```

```
49      // Create std::vector with our allocator.
50      std::vector<int,
51          libmemkind::pmem::allocator<int>> v(alc);
52
53      for (int i = 0; i < 100; ++i)
54          v.push_back(i);
55
56      for (int i = 0; i < 100; ++i)
57          assert(v[i] == i);
```

- Line 43: We define a persistent memory pool of 64MiB.

- Lines 46-47: We create an allocator object alc of type
  pmem::allocator<int>.

- Line 50: We create a vector object v of type std::vector<int,
  pmem::allocator<int> > and pass in the alc from line 47 object as
  an argument. The pmem::allocator is stateful and has no default
  constructor. This requires passing the allocator object to the vector
  constructor; otherwise, a compilation error occurs if the default
  constructor of std::vector<int, pmem::allocator<int> > is called
  because the vector constructor will try to call the default constructor
  of pmem::allocator, which does not exist yet.

Creating a Vector of Strings

Listing 10-11 shows how to create a vector of strings that resides in persistent memory.
We define pmem_string as a typedef of std::basic_string with pmem::allocator.
In this example, std::scoped_allocator_adaptor allows the vector to propagate the
pmem::allocator instance to all pmem_string objects stored in the vector object.

Listing 10-11. vector_of_strings.cpp: creating a vector of strings

```
37   #include <pmem_allocator.h>
38   #include <vector>
39   #include <string>
40   #include <scoped_allocator>
41   #include <cassert>
```

```
42   #include <iostream>
43
44   typedef libmemkind::pmem::allocator<char> str_alloc_type;
45
46   typedef std::basic_string<char, std::char_traits<char>,
     str_alloc_type> pmem_string;
47
48   typedef libmemkind::pmem::allocator<pmem_string> vec_alloc_type;
49
50   typedef std::vector<pmem_string, std::scoped_allocator_adaptor
     <vec_alloc_type> > vector_type;
51
52   int main(int argc, char *argv[]) {
53       const size_t pmem_max_size = 64 * 1024 * 1024; //64 MB
54       const std::string pmem_dir("/daxfs");
55
56       // Create allocator object
57       vec_alloc_type alc(pmem_dir, pmem_max_size);
58       // Create std::vector with our allocator.
59       vector_type v(alc);
60
61       v.emplace_back("Foo");
62       v.emplace_back("Bar");
63
64       for (auto str : v) {
65               std::cout << str << std::endl;
66       }
```

- Line 46: We define pmem_string as a typedef of std::basic_string.

- Line 48: We define the pmem::allocator using the pmem_string type.

- Line 50: Using std::scoped_allocator_adaptor allows the vector to propagate the pmem::allocator instance to all pmem_string objects stored in the vector object.

Expanding Volatile Memory Using Persistent Memory

Persistent memory is treated by the kernel as a device. In a typical use-case, a persistent memory-aware file system is created and mounted with the *–o dax* option, and files are memory-mapped into the virtual address space of a process to give the application direct load/store access to persistent memory regions.

A new feature was added to the Linux kernel v5.1 such that persistent memory can be used more broadly as volatile memory. This is done by binding a persistent memory device to the kernel, and the kernel manages it as an extension to DRAM. Since persistent memory has different characteristics than DRAM, memory provided by this device is visible as a separate NUMA node on its corresponding socket.

To use the MEMKIND_DAX_KMEM kind, you need pmem to be available using *device DAX*, which exposes pmem as devices with names like /dev/dax*. If you have an existing dax device and want to migrate the device model type to use DEV_DAX_KMEM, use:

```
$ sudo daxctl migrate-device-model
```

To create a new dax device using all available capacity on the first available region (NUMA node), use:

```
$ sudo ndctl create-namespace --mode=devdax --map=mem
```

To create a new dax device specifying the region and capacity, use:

```
$ sudo ndctl create-namespace --mode=devdax --map=mem --region=region0
--size=32g
```

To display a list of namespaces, use:

```
$ ndctl list
```

If you have already created a namespace in another mode, such as the default fsdax, you can reconfigure the device using the following where namespace0.0 is the existing namespace you want to reconfigure:

```
$ sudo ndctl create-namespace --mode=devdax --map=mem --force -e namespace0.0
```

For more details about creating new namespace read https://docs.pmem.io/ndctl-users-guide/managing-namespaces#creating-namespaces.

DAX devices must be converted to use the `system-ram` mode. Converting a dax device to a NUMA node suitable for use with system memory can be performed using following command:

```
$ sudo daxctl reconfigure-device dax2.0 --mode=system-ram
```

This will migrate the device from using the device_dax driver to the dax_pmem driver. The following shows an example output with dax1.0 configured as the default devdax type and `dax2.0` is `system-ram`:

```
$ daxctl list
    [
      {
        "chardev":"dax1.0",
        "size":263182090240,
        "target_node":3,
        "mode":"devdax"
      },
      {
        "chardev":"dax2.0",
        "size":263182090240,
        "target_node":4,
        "mode":"system-ram"
      }
    ]
```

You can now use numactl -H to show the hardware NUMA configuration. The following example output is collected from a 2-socket system and shows node 4 is a new system-ram backed NUMA node created from persistent memory:

```
$ numactl -H
    available: 3 nodes (0-1,4)
    node 0 cpus: 0 1 2 3 4 5 6 7 8 9 10 11 12 13 14 15 16 17 18 19 20 21 22
                 23 24 25 26 27 56 57 58 59 60 61 62 63 64 65 66 67 68 69
                 70 71 72 73 74 75 76 77 78 79 80 81 82 83
    node 0 size: 192112 MB
    node 0 free: 185575 MB
```

```
node 1 cpus: 28 29 30 31 32 33 34 35 36 37 38 39 40 41 42 43 44 45 46
             47 48 49 50 51 52 53 54 55 84 85 86 87 88 89 90 91 92 93
             94 95 96 97 98 99 100 101 102 103 104 105 106 107 108 109
             110 111
node 1 size: 193522 MB
node 1 free: 193107 MB
node 4 cpus:
node 4 size: 250880 MB
node 4 free: 250879 MB
node distances:
node   0   1   4
  0:  10  21  17
  1:  21  10  28
  4:  17  28  10
```

To online the NUMA node and have the Kernel manage the new memory, use:

```
$ sudo daxctl online-memory dax0.1
dax0.1: 5 sections already online
dax0.1: 0 new sections onlined
onlined memory for 1 device
```

At this point, the kernel will use the new capacity for normal operation. The new memory shows itself in tools such lsmem example shown below where we see an additional 10GiB of system-ram in the 0x0000003380000000-0x00000035ffffffff address range:

```
$ lsmem
RANGE                                     SIZE  STATE  REMOVABLE   BLOCK
0x0000000000000000-0x000000007fffffff       2G online         no       0
0x0000000100000000-0x000000277fffffff     154G online        yes    2-78
0x0000002780000000-0x000000297fffffff       8G online         no   79-82
0x0000002980000000-0x0000002effffffff      22G online        yes   83-93
0x0000002f00000000-0x0000002fffffffff       4G online         no   94-95
0x0000003380000000-0x00000035ffffffff      10G online        yes 103-107
0x000001aa80000000-0x000001d0ffffffff     154G online        yes 853-929
0x000001d100000000-0x000001d37fffffff      10G online         no 930-934
0x000001d380000000-0x000001d8ffffffff      22G online        yes 935-945
0x000001d900000000-0x000001d9ffffffff       4G online         no 946-947
```

```
Memory block size:          2G
Total online memory:      390G
Total offline memory:       0B
```

To programmatically allocate memory from a NUMA node created using persistent memory, a new static kind, called MEMKIND_DAX_KMEM, was added to `libmemkind` that uses the `system-ram` DAX device.

Using `MEMKIND_DAX_KMEM` as the first argument to memkind_malloc(), shown below, you can use persistent memory from separate NUMA nodes in a single application. The persistent memory is still physically connected to a CPU socket, so the application should take care to ensure CPU affinity for optimal performance.

```
memkind_malloc(MEMKIND_DAX_KMEM, size_t size)
```

Figure 10-3 shows an application that created two static kind objects: `MEMKIND_DEFAULT` and `MEMKIND_DAX_KMEM`.

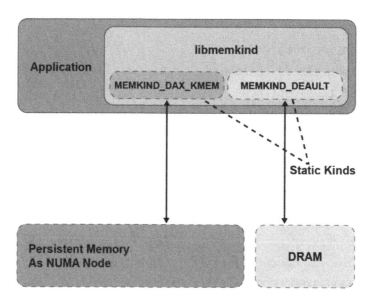

Figure 10-3. *An application that created two kind objects from different types of memory*

The difference between the PMEM_KIND described earlier and MEMKIND_DAX_KMEM is that the MEMKIND_DAX_KMEM is a static kind and uses mmap() with the MAP_PRIVATE flag, while the dynamic PMEM_KIND is created with memkind_create_pmem() and uses the MAP_SHARED flag when memory-mapping files on a DAX-enabled file system.

Child processes created using the `fork(2)` system call inherit the `MAP_PRIVATE` mappings from the parent process. When memory pages are modified by the parent process, a copy-on-write mechanism is triggered by the kernel to create an unmodified copy for the child process. These pages are allocated on the same NUMA node as the original page.

libvmemcache: An Efficient Volatile Key-Value Cache for Large-Capacity Persistent Memory

Some existing in-memory databases (IMDB) rely on manual dynamic memory allocations (`malloc`, `jemalloc`, `tcmalloc`), which can exhibit external and internal memory fragmentation when run for a long period of time, leaving large amounts of memory un-allocatable. Internal and external fragmentation is briefly explained as follows:

- *Internal fragmentation* occurs when more memory is allocated than is required, and the unused memory is contained within the allocated region. For example, if the requested allocation size is 200 bytes, a chunk of 256 bytes is allocated.

- *External fragmentation* occurs when variable memory sizes are allocated dynamically, resulting in a failure to allocate a contiguous chunk of memory, although the requested chunk of memory remains available in the system. This problem is more pronounced when large capacities of persistent memory are being used as volatile memory. Applications with substantially long runtimes need to solve this problem, especially if the allocated sizes have considerable variation. Applications and runtime environments handle this problem in different ways, for example:

  - Java and .NET use compacting garbage collection

  - Redis and Apache Ignite* use defragmentation algorithms

  - Memcached uses a slab allocator

Each of the above allocator mechanisms has pros and cons. Garbage collection and defragmentation algorithms require processing to occur on the heap to free unused allocations or move data to create contiguous space. Slab allocators usually define a fixed set of different sized buckets at initialization without knowing how many of each bucket the application will need. If the slab allocator depletes a certain bucket size, it allocates from larger sized buckets, which reduces the amount of free space. These mechanisms can potentially block the application's processing and reduce its performance.

libvmemcache Overview

`libvmemcache` is an embeddable and lightweight in-memory caching solution with a key-value store at its core. It is designed to take full advantage of large-capacity memory, such as persistent memory, efficiently using memory mapping in a scalable way. It is optimized for use with memory-addressable persistent storage through a DAX-enabled file system that supports load/store operations. `libvmemcache` has these unique characteristics:

- The extent-based memory allocator sidesteps the fragmentation problem that affects most in-memory databases, and it allows the cache to achieve very high space utilization for most workloads.

- Buffered LRU (least recently used) combines a traditional LRU doubly linked list with a non-blocking ring buffer to deliver high scalability on modern multicore CPUs.

- A unique indexing `critnib` data structure delivers high performance and is very space efficient.

The cache for `libvmemcache` is tuned to work optimally with relatively large value sizes. While the smallest possible size is 256 bytes, `libvmemcache` performs best if the expected value sizes are above 1 kilobyte.

`libvmemcache` has more control over the allocation because it implements a custom memory-allocation scheme using an extents-based approach (like that of file system extents). `libvmemcache` can, therefore, concatenate and achieve substantial space efficiency. Additionally, because it is a cache, it can evict data to allocate new entries in a worst-case scenario. `libvmemcache` will *always* allocate exactly as much memory as it freed, minus metadata overhead. This is not true for caches based on common memory allocators such as memkind. `libvmemcache` is designed to work with terabyte-sized in-memory workloads, with very high space utilization.

libvmemcache works by automatically creating a temporary file on a DAX-enabled file system and memory-mapping it into the application's virtual address space. The temporary file is deleted when the program terminates and gives the perception of volatility. Figure 10-4 shows the application using traditional malloc() to allocate memory from DRAM and using libvmemcache to memory map a temporary file residing on a DAX-enabled file system from persistent memory.

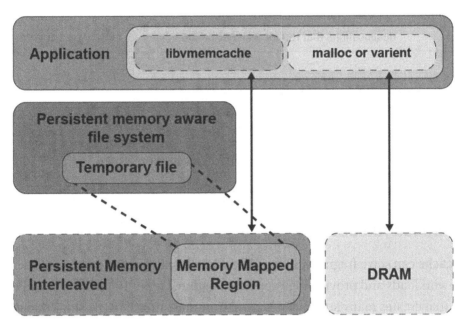

Figure 10-4. *An application using libvmemcache memory-maps a temporary file from a DAX-enabled file system*

Although libmemkind supports different kinds of memory and memory consumption policies, the underlying allocator is jemalloc, which uses dynamic memory allocation. Table 10-2 compares the implementation details of libvmemcache and libmemkind.

Table 10-2. *Design aspects of libmemkind and libvmemcache*

	libmemkind (PMEM)	**libvmemcache**
Allocation Scheme	Dynamic allocator	Extent based (not restricted to sector, page, etc.)
Purpose	General purpose	Lightweight in-memory cache
Fragmentation	Apps with random size allocations/ deallocations that run for a longer period	Minimized

libvmemcache Design

libvmemcache has two main design aspects:

1. Allocator design to improve/resolve fragmentation issues

2. A scalable and efficient LRU policy

Extent-Based Allocator

libvmemcache can solve fragmentation issues when working with terabyte-sized in-memory workloads and provide high space utilization. Figure 10-5 shows a workload example that creates many small objects, and over time, the allocator stops due to fragmentation.

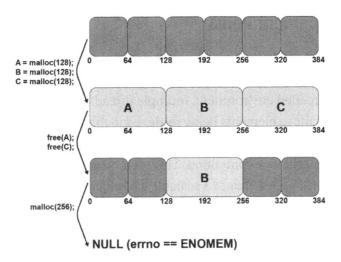

Figure 10-5. *An example of a workload that creates many small objects, and the allocator stops due to fragmentation*

libvmemcache uses an extent-based allocator, where an extent is a contiguous set of blocks allocated for storing the data in a database. Extents are typically used with large blocks supported by file systems (sectors, pages, etc.), but such restrictions do not apply when working with persistent memory that supports smaller block sizes (cache line). Figure 10-6 shows that if a single contiguous free block is not available to allocate an object, multiple, noncontiguous blocks are used to satisfy the allocation request. The noncontiguous allocations appear as a single allocation to the application.

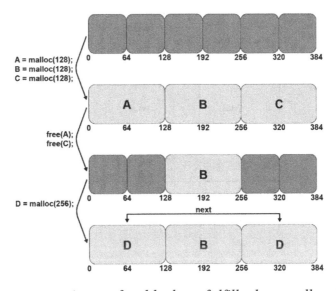

Figure 10-6. *Using noncontiguous free blocks to fulfill a larger allocation request*

Scalable Replacement Policy

An LRU cache is traditionally implemented as a doubly linked list. When an item is retrieved from this list, it gets moved from the middle to the front of the list, so it is not evicted. In a multithreaded environment, multiple threads may contend with the front element, all trying to move elements being retrieved to the front. Therefore, the front element is always locked (along with other locks) before moving the element being retrieved, which results in lock contention. This method is not scalable and is inefficient.

A buffer-based LRU policy creates a scalable and efficient replacement policy. A non-blocking ring buffer is placed in front of the LRU linked list to track the elements being retrieved. When an element is retrieved, it is added to this buffer, and only when the buffer is full (or the element is being evicted), the linked list is locked, and the elements in that buffer are processed and moved to the front of the list. This method preserves the LRU policy and provides a scalable LRU mechanism with minimal performance impact. Figure 10-7 shows a ring buffer-based design for the LRU algorithm.

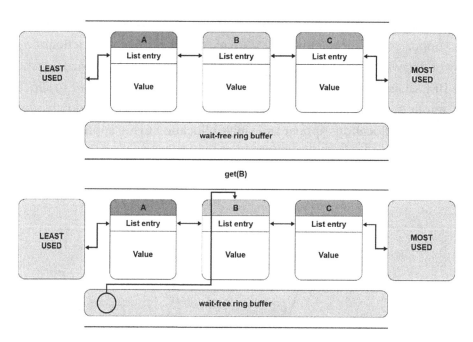

Figure 10-7. *A ring buffer-based LRU design*

Using libvmemcache

Table 10-3 lists the basic functions that libvmemcache provides. For a complete list, see the libvmemcache man pages (https://pmem.io/vmemcache/manpages/master/vmemcache.3.html).

Table 10-3. *The libvmemcache functions*

Function Name	Description
vmemcache_new	Creates an empty unconfigured vmemcache instance with default values: Eviction_policy=VMEMCACHE_REPLACEMENT_LRU Extent_size = VMEMCAHE_MIN_EXTENT VMEMCACHE_MIN_POOL
vmemcache_add	Associates the cache with a path.
vmemcache_set_size	Sets the size of the cache.
vmemcache_set_extent_size	Sets the block size of the cache (256 bytes minimum).
vmemcache_set_eviction_policy	Sets the eviction policy: 1. VMEMCACHE_REPLACEMENT_NONE 2. VMEMCACHE_REPLACEMENT_LRU
vmemcache_add	Associates the cache with a given path on a DAX-enabled file system or non-DAX-enabled file system.
vmemcache_delete	Frees any structures associated with the cache.
vmemcache_get	Searches for an entry with the given key, and if found, the entry's value is copied to vbuf.
vmemcache_put	Inserts the given key-value pair into the cache.
vmemcache_evict	Removes the given key from the cache.
vmemcache_callback_on_evict	Called when an entry is being removed from the cache.
vmemcache_callback_on_miss	Called when a get query fails to provide an opportunity to insert the missing key.

To illustrate how libvmemcache is used, Listing 10-12 shows how to create an instance of vmemcache using default values. This example uses a temporary file on a DAX-enabled file system and shows how a callback is registered after a cache miss for a key "meow."

Listing 10-12. vmemcache.c: An example program using libvmemcache

```
37  #include <libvmemcache.h>
38  #include <stdio.h>
39  #include <stdlib.h>
40  #include <string.h>
41
42  #define STR_AND_LEN(x) (x), strlen(x)
43
44  VMEMcache *cache;
45
46  void on_miss(VMEMcache *cache, const void *key,
47      size_t key_size, void *arg)
48  {
49      vmemcache_put(cache, STR_AND_LEN("meow"),
50          STR_AND_LEN("Cthulhu fthagn"));
51  }
52
53  void get(const char *key)
54  {
55      char buf[128];
56      ssize_t len = vmemcache_get(cache,
57      STR_AND_LEN(key), buf, sizeof(buf), 0, NULL);
58      if (len >= 0)
59          printf("%.*s\n", (int)len, buf);
60      else
61          printf("(key not found: %s)\n", key);
62  }
63
64  int main()
65  {
```

```
66    cache = vmemcache_new();
67    if (vmemcache_add(cache, "/daxfs")) {
68        fprintf(stderr, "error: vmemcache_add: %s\n",
69                vmemcache_errormsg());
70            exit(1);
71    }
72
73    // Query a non-existent key
74    get("meow");
75
76    // Insert then query
77    vmemcache_put(cache, STR_AND_LEN("bark"),
78        STR_AND_LEN("Lorem ipsum"));
79    get("bark");
80
81    // Install an on-miss handler
82    vmemcache_callback_on_miss(cache, on_miss, 0);
83    get("meow");
84
85    vmemcache_delete(cache);
```

- Line 66: Creates a new instance of vmemcache with default values for eviction_policy and extent_size.

- Line 67: Calls the vmemcache_add() function to associate cache with a given path.

- Line 74: Calls the get() function to query on an existing key. This function calls the vmemcache_get() function with error checking for success/failure of the function.

- Line 77: Calls vmemcache_put() to insert a new key.

- Line 82: Adds an on-miss callback handler to insert the key "meow" into the cache.

- Line 83: Retrieves the key "meow" using the get() function.

- Line 85: Deletes the vmemcache instance.

Summary

This chapter showed how persistent memory's large capacity can be used to hold volatile application data. Applications can choose to allocate and access data from DRAM or persistent memory or both.

memkind is a very flexible and easy-to-use library with semantics that are similar to the `libc malloc/free` APIs that developers frequently use.

`libvmemcache` is an embeddable and lightweight in-memory caching solution that allows applications to efficiently use persistent memory's large capacity in a scalable way. `libvmemcache` is an open source project available on GitHub at `https://github.com/pmem/vmemcache`.

## CHAPTER 11

Designing Data Structures for Persistent Memory

Taking advantage of the unique characteristics of persistent memory, such as byte addressability, persistence, and update in place, allows us to build data structures that are much faster than any data structure requiring serialization or flushing to a disk. However, this comes at a cost. Algorithms must be carefully designed to properly persist data by flushing CPU caches or using non-temporal stores and memory barriers to maintain data consistency. This chapter describes how to design such data structures and algorithms and shows what properties they should have.

Contiguous Data Structures and Fragmentation

Fragmentation is one of the most critical factors to consider when designing a data structure for persistent memory due to the length of heap life. A persistent heap can live for years with different versions of an application. In volatile use cases, the heap is destroyed when the application exits. The life of the heap is usually measured in hours, days, or weeks.

Using file-backed pages for memory allocation makes it difficult to take advantage of the operating system–provided mechanisms for minimizing fragmentation, such as presenting discontinuous physical memory as a contiguous virtual region. It is possible to manually manage virtual memory at a low granularity, producing a page-level defragmentation mechanism for objects in user space. But this mechanism could lead to complete fragmentation of physical memory and an inability to take advantage of huge pages. This can cause an increased number of translation lookaside buffer (TLB) misses, which significantly slows down the entire application. To make effective use of persistent memory, you should design data structures in a way that minimizes fragmentation.

© The Author(s) 2020
S. Scargall, *Programming Persistent Memory*, https://doi.org/10.1007/978-1-4842-4932-1_11

Internal and External Fragmentation

Internal fragmentation refers to space that is overprovisioned inside allocated blocks. An allocator always returns memory in fixed-sized *chunks* or *buckets*. The allocator must determine what size each bucket is and how many different sized buckets it provides. If the size of the memory allocation request does not exactly match a predefined bucket size, the allocator will return a larger memory bucket. For example, if the application requests a memory allocation of 200KiB, but the allocator has bucket sizes of 128KiB and 256KiB, the request is allocated from an available 256KiB bucket. The allocator must usually return a memory chunk with a size divisible by 16 due to its internal alignment requirements.

External fragmentation occurs when free memory is scattered in small blocks. For example, imagine using up the entire memory with 4KiB allocations. If we then free every other allocation, we have half of the memory available; however, we cannot allocate more than 4KiB at once because that is the maximum size of any contiguous free space. Figure 11-1 illustrates this fragmentation, where the red cells represent allocated space and the white cells represent free space.

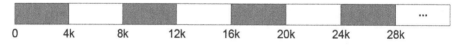

Figure 11-1. *External fragmentation*

When storing a sequence of elements in persistent memory, several possible data structures can be used:

- Linked list: Each node is allocated from persistent memory.

- Dynamic array (vector): A data structure that pre-allocates memory in bigger chunks. If there is no free space for new elements, it allocates a new array with bigger capacity and moves all elements from the old array to the new one.

- Segment vector: A list of fixed-size arrays. If there is no free space left in any segment, a new one is allocated.

Consider fragmentation for each of those data structures:

- For linked lists, fragmentation efficiency depends on the node size. If it is small enough, then high internal fragmentation can be expected. During node allocation, every allocator will return memory with a certain alignment that will likely be different than the node size.

- Using dynamic array results in fewer memory allocations, but every allocation will have a different size (most implementations double the previous one), which results in a higher external fragmentation.

- Using a segment vector, the size of a segment is fixed, so every allocation has the same size. This practically eliminates external fragmentation because we can allocate a new one for each freed segment.[1]

Atomicity and Consistency

Guaranteeing consistency requires the proper ordering of stores and making sure data is stored persistently. To make an atomic store bigger than 8 bytes, you must use some additional mechanisms. This section describes several mechanisms and discusses their memory and time overheads. For the time overhead, the focus is on analyzing the number of flushes and memory barriers used because they have the biggest impact on performance.

Transactions

One way to guarantee atomicity and consistency is to use transactions (described in detail in Chapter 7). Here we focus on how to design a data structure to use transactions efficiently. An example data structure that uses transactions is described in the "Sorted Array with Versioning" section later in this chapter.

Transactions are the simplest solution for guaranteeing consistency. While using transactions can easily make most operations atomic, two items must be kept in mind. First, transactions that use logging always introduce memory and time overheads. Second, in the case of undo logging, the memory overhead is proportional to the size of data you modify, while the time overhead depends on the number of snapshots. Each snapshot must be persisted prior to the modification of snapshotted data.

---

[1]Using the `libpmemobj` allocator, it is also possible to easily lower internal fragmentation by using allocation classes (see Chapter 7).

It is recommended to use a data-oriented approach when designing a data structure for persistent memory. The idea is to store data in such a way that its processing by the CPU is cache friendly. Imagine having to store a sequence of 1000 records that consist of 2 integer values. This has two approaches: Either use two arrays of integers as shown in Listing 11-1, or use one array of pairs as shown in Listing 11-2. The first approach is SoA (Structure of Arrays), and the second is AoS (Array of Structures).

Listing 11-1. SoA layout approach to store data

```
struct soa {
    int a[1000];
    int b[1000];
};
```

Listing 11-2. AoS layout approach to store data

```
std::pair<int, int> aos_records[1000];
```

Depending on the access pattern to the data, you may prefer one solution over the other. If the program frequently updates both fields of an element, then the AoS solution is better. However, if the program only updates the first variable of all elements, then the SoA solution works best.

For applications that use volatile memory, the main concerns are usually cache misses and optimizations for single instruction, multiple data (SIMD) processing. SIMD is a class of parallel computers in Flynn's taxonomy,[2] which describes computers with multiple processing elements that simultaneously perform the same operation on multiple data points. Such machines exploit data-level parallelism, but not concurrency: There are simultaneous (parallel) computations but only a single process (instruction) at a given moment.

While those are still valid concerns for persistent memory, developers must consider snapshotting performance when transactions are used. Snapshotting one contiguous memory region is always better then snapshotting several smaller regions, mainly due to the smaller overhead incurred by using less metadata. Efficient data structure layout that takes these considerations into account is imperative for avoiding future problems when migrating data from DRAM-based implementations to persistent memory.

---

[2]For a full definition of SIMD, see https://en.wikipedia.org/wiki/SIMD.

Listing 11-3 presents both approaches; in this example, we want to increase the first integer by one.

Listing 11-3. Layout and snapshotting performance

```
37 struct soa {
38   int a[1000];
39   int b[1000];
40 };
41
42 struct root {
43   soa soa_records;
44   std::pair<int, int aos_records[1000];
45 };
46
47 int main()
48 {
49   try {
50     auto pop = pmem::obj::pool<root>::create("/daxfs/pmpool",
51             "data_oriented", PMEMOBJ_MIN_POOL, 0666);
52
53   auto root = pop.root();
54
55   pmem::obj::transaction::run(pop, [&]{
56     pmem::obj::transaction::snapshot(&root->soa_records);
57     for (int i = 0; i < 1000; i++) {
58       root->soa_records.a[i]++;
59     }
60
61     for (int i = 0; i < 1000; i++) {
62       pmem::obj::transaction::snapshot(
63                     &root->aos_records[i].first);
64       root->aos_records[i].first++;
65     }
66   });
67
```

```
68   pop.close();
69   } catch (std::exception &e) {
70     std::cerr << e.what() << std::endl;
71   }
72 }
```

- Lines 37-45: We define two different data structures to store records of integers. The first one is SoA – where we store integers in two separate arrays. Line 44 shows a single array of pairs – AoS.

- Lines 56-59: We take advantage of the SoA layout by snapshotting the entire array at once. Then we can safely modify each element.

- Lines 61-65: When using AoS, we are forced to snapshot data in every iteration – elements we want to modify are not contiguous in memory.

Examples of data structures that use transactions are shown in the "Hash Table with Transactions" and "Hash Table with Transactions and Selective Persistence" sections, later in this chapter.

Copy-on-Write and Versioning

Another way to maintain consistency is the copy-on-write (CoW) technique. In this approach, every modification creates a new version at a new location whenever you want to modify some part of a persistent data structure. For example, a node in a linked list can use the CoW approach as described in the following:

1. Create a copy of the element in the list. If a copy is dynamically allocated in persistent memory, you should also save the pointer in persistent memory to avoid a memory leak. If you fail to do that and the application crashes after the allocation, then on the application restart, newly allocated memory will be unreachable.

2. Modify the copy and persist the changes.

3. Atomically change the original element with the copy and persist the changes, then free the original node if needed. After this step successfully completes, the element is updated and is in a consistent state. If a crash occurs before this step, the original element is untouched.

Although using this approach compared to transactions can be faster, it is significantly harder to implement because you must manually persist data.

Copy-on-write usually works well in multithreaded systems where mechanisms like reference counting or garbage collection are used to free copies that are no longer used. Although such systems are beyond the scope of this book, Chapter 14 describes concurrency in multithreaded applications.

Versioning is a very similar concept to copy-on-write. The difference is that here you hold more than one version of a data field. Each modification creates a new version of the field and stores information about the current one. The example presented in "Sorted Array with Versioning" later in this chapter shows this technique in an implementation of the insert operation for a sorted array. In the preceding example, only two versions of a variable are kept, the old and current one as a two-element array. The insert operations alternately write data to the first and second element of this array.

Selective Persistence

Persistent memory is faster than disk storage but potentially slower than DRAM. Hybrid data structures, where some parts are stored in DRAM and some parts are in persistent memory, can be implemented to accelerate performance. Caching previously computed values or frequently accessed parts of a data structure in DRAM can improve access latency and improve overall performance.

Data does not always need to be stored in persistent memory. Instead, it can be rebuilt during the restart of an application to provide a performance improvement during runtime given that it accesses data from DRAM and does not require transactions. An example of this approach appears in "Hash Table with Transactions and Selective Persistence."

Example Data Structures

This section presents several data structure examples that were designed using the previously described methods for guaranteeing consistency. The code is written in C++ and uses `libpmemobj-cpp`. See Chapter 8 for more information about this library.

Hash Table with Transactions

We present an example of a hash table implemented using transactions and containers using `libpmemobj-cpp`.

As a quick primer to some, and a refresher to other readers, a hash table is a data structure that maps keys to values and guarantees O(1) lookup time. It is usually implemented as an array of buckets (a bucket is a data structure that can hold one or more key-value pairs). When inserting a new element to the hash table, a hash function is applied to the element's key. The resulting value is treated as an index of a bucket to which the element is inserted. It is possible that the result of the hash function for different keys will be the same; this is called a *collision*. One method for resolving collisions is to use separate chaining. This approach stores multiple key-value pairs in one bucket; the example in Listing 11-4 uses this method.

For simplicity, the hash table in Listing 11-4 only provides the `const Value& get(const std::string &key)` and `void put(const std::string &key, const Value &value)` methods. It also has a fixed number of buckets. Extending this data structure to support the remove operation and to have a dynamic number of buckets is left as an exercise to you.

Listing 11-4. Implementation of a hash table using transactions

```
38    #include <functional>
39    #include <libpmemobj++/p.hpp>
40    #include <libpmemobj++/persistent_ptr.hpp>
41    #include <libpmemobj++/pext.hpp>
42    #include <libpmemobj++/pool.hpp>
43    #include <libpmemobj++/transaction.hpp>
44    #include <libpmemobj++/utils.hpp>
45    #include <stdexcept>
46    #include <string>
47
48    #include "libpmemobj++/array.hpp"
49    #include "libpmemobj++/string.hpp"
50    #include "libpmemobj++/vector.hpp"
51
```

```
52   /**
53    * Value - type of the value stored in hashmap
54    * N - number of buckets in hashmap
55    */
56   template <typename Value, std::size_t N>
57   class simple_kv {
58   private:
59     using key_type = pmem::obj::string;
60     using bucket_type = pmem::obj::vector<
61         std::pair<key_type, std::size_t>>;
62     using bucket_array_type = pmem::obj::array<bucket_type, N>;
63     using value_vector = pmem::obj::vector<Value>;
64
65     bucket_array_type buckets;
66     value_vector values;
67
68   public:
69     simple_kv() = default;
70
71     const Value &
72     get(const std::string &key) const
73     {
74     auto index = std::hash<std::string>{}(key) % N;
75
76     for (const auto &e : buckets[index]) {
77      if (e.first == key)
78        return values[e.second];
79     }
80
81     throw std::out_of_range("no entry in simplekv");
82     }
83
```

```
84   void
85   put(const std::string &key, const Value &val)
86   {
87    auto index = std::hash<std::string>{}(key) % N;
88
89    /* get pool on which this simple_kv resides */
90    auto pop = pmem::obj::pool_by_vptr(this);
91
92    /* search for element with specified key - if found
93     * update its value in a transaction*/
94    for (const auto &e : buckets[index]) {
95      if (e.first == key) {
96        pmem::obj::transaction::run(
97          pop, [&] { values[e.second] = val; });
98
99        return;
100       }
101     }
102
103    /* if there is no element with specified key, insert
104     * new value to the end of values vector and put
105     * reference in proper bucket */
106    pmem::obj::transaction::run(pop, [&] {
107      values.emplace_back(val);
108      buckets[index].emplace_back(key, values.size() - 1);
109      });
110     }
111   };
```

- Lines 58-66: Define the layout of a hash map as a pmem::obj::array
  of buckets, where each bucket is a pmem::obj::vector of key and
  index pairs and pmem::obj::vector contains the values. The index
  in a bucket entry always specifies a position of the actual value
  stored in a separate vector. For snapshotting optimization, the value
  is not saved next to a key in a bucket. When obtaining a non-const
  reference to an element in pmem::obj::vector, the element is always

snapshotted. To avoid snapshotting unnecessary data, for example, if the key is immutable, we split keys and values into separate vectors. This also helps in the case of updating several values in one transaction. Recall the discussion in the "Copy-on-Write and Versioning" section. The result could turn out to be next to each other in a vector, and there could be fewer bigger regions to snapshot.

- Line 74: Calculate hash in a table using standard library feature.

- Lines 76-79: Search for entry with specified key by iterating over all buckets stored in the table under `index`. Note that e is a const reference to the key-value pair. Because of the way `libpmemobj-cpp` containers work, this has a positive impact on performance when compared to non-const reference; obtaining non-const reference requires a snapshot, while a const reference does not.

- Line 90: Get the instance of the `pmemobj` pool object, which is used to manage the persistent memory pool where our data structure resides.

- Lines 94-95: Find the position of a value in the values vector by iterating over all the entries in the designated bucket.

- Lines 96-98: If an element with the specified key is found, update its value using a transaction.

- Lines 106-109: If there is no element with the specified key, insert a value into the values vector, and put a reference to this value in the proper bucket; that is, create key, index pair. Those two operations must be completed in a single atomic transaction because we want them both to either succeed or fail.

Hash Table with Transactions and Selective Persistence

This example shows how to modify a persistent data structure (hash table) by moving some data out of persistent memory. The data structure presented in Listing 11-5 is a modified version of the hash table in Listing 11-4 and contains the implementation of this hash table design. Here we store only the vector of keys and vector of values in persistent memory. On application startup, we build the buckets and store them in volatile memory for faster processing during runtime. The most noticeable performance gain would be in the `get()` method.

Listing 11-5. Implementation of hash table with transactions and selective persistence

```
40 #include <array>
41 #include <functional>
42 #include <libpmemobj++/p.hpp>
43 #include <libpmemobj++/persistent_ptr.hpp>
44 #include <libpmemobj++/pext.hpp>
45 #include <libpmemobj++/pool.hpp>
46 #include <libpmemobj++/transaction.hpp>
47 #include <libpmemobj++/utils.hpp>
48 #include <stdexcept>
49 #include <string>
50 #include <vector>
51
52 #include "libpmemobj++/array.hpp"
53 #include "libpmemobj++/string.hpp"
54 #include "libpmemobj++/vector.hpp"
55
56 template <typename Value, std::size_t N>
57 struct simple_kv_persistent;
58
59 /**
60  * This class is runtime wrapper for simple_kv_peristent.
61  * Value - type of the value stored in hashmap
62  * N - number of buckets in hashmap
63  */
64 template <typename Value, std::size_t N>
65 class simple_kv_runtime {
66 private:
67   using volatile_key_type = std::string;
68   using bucket_entry_type = std::pair<volatile_key_type, std::size_t>;
69   using bucket_type = std::vector<bucket_entry_type>;
70   using bucket_array_type = std::array<bucket_type, N>;
71
```

```cpp
72   bucket_array_type buckets;
73   simple_kv_persistent<Value, N> *data;
74
75 public:
76   simple_kv_runtime(simple_kv_persistent<Value, N> *data)
77   {
78    this->data = data;
79
80    for (std::size_t i = 0; i < data->values.size(); i++) {
81     auto volatile_key = std::string(data->keys[i].c_str(),
82                 data->keys[i].size());
83
84     auto index = std::hash<std::string>{}(volatile_key)%N;
85     buckets[index].emplace_back(
86      bucket_entry_type{volatile_key, i});
87    }
88   }
89
90   const Value &
91   get(const std::string &key) const
92   {
93    auto index = std::hash<std::string>{}(key) % N;
94
95    for (const auto &e : buckets[index]) {
96     if (e.first == key)
97       return data->values[e.second];
98    }
99
100   throw std::out_of_range("no entry in simplekv");
101  }
102
103  void
104  put(const std::string &key, const Value &val)
105  {
106   auto index = std::hash<std::string>{}(key) % N;
107
```

```
108    /* get pool on which persistent data resides */
109      auto pop = pmem::obj::pool_by_vptr(data);
110
111    /* search for element with specified key - if found
112     * update its value in a transaction */
113    for (const auto &e : buckets[index]) {
114    if (e.first == key) {
115      pmem::obj::transaction::run(pop, [&] {
116        data->values[e.second] = val;
117      });
118
119      return;
120     }
121    }
122
123    /* if there is no element with specified key, insert new value
124     * to the end of values vector and key to keys vector
125     * in a transaction */
126    pmem::obj::transaction::run(pop, [&] {
127     data->values.emplace_back(val);
128     data->keys.emplace_back(key);
129    });
130
131    buckets[index].emplace_back(key, data->values.size() - 1);
132  }
133 };
134
135 /**
136  * Class which is stored on persistent memory.
137  * Value - type of the value stored in hashmap
138  * N - number of buckets in hashmap
139  */
140 template <typename Value, std::size_t N>
141 struct simple_kv_persistent {
142  using key_type = pmem::obj::string;
```

```
143  using value_vector = pmem::obj::vector<Value>;
144  using key_vector = pmem::obj::vector<key_type>;
145
146 /* values and keys are stored in separate vectors to optimize
147  * snapshotting. If they were stored as a pair in single vector
148  * entire pair would have to be snapshotted in case of value update */
149  value_vector values;
150  key_vector keys;
151
152  simple_kv_runtime<Value, N>
153  get_runtime()
154  {
155    return simple_kv_runtime<Value, N>(this);
156  }
157 };
```

- Line 67: We define the data types residing in volatile memory. These are very similar to the types used in the persistent version in "Hash Table with Transactions." The only difference is that here we use std containers instead of pmem::obj.

- Line 72: We declare the volatile buckets array.

- Line 73: We declare the pointer to persistent data (simple_kv_persistent structure).

- Lines 75-88: In the simple_kv_runtime constructor, we rebuild the bucket's array by iterating over keys and values in persistent memory. In volatile memory, we store both the keys, which are a copy of the persistent data and the index for the values vector in persistent memory.

- Lines 90-101: The get() function looks for an element reference in the volatile buckets array. There is only one reference to persistent memory when we read the actual value on line 97.

- Lines 113-121: Similar to the get() function, we search for an element using the volatile data structure and, when found, update the value in a transaction.

- Lines 126-129: When there is no element with the specified key in the hash table, we insert both a value and a key to their respective vectors in persistent memory in a transaction.

- Line 131: After inserting data to persistent memory, we update the state of the volatile data structure. Note that this operation does not have to be atomic. If a program crashes, the bucket array will be rebuilt on startup.

- Lines 149-150: We define the layout of the persistent data. Key and values are stored in separate `pmem::obj::vector`.

- Lines 153-156: We define a function that returns the runtime object of this hash table.

Sorted Array with Versioning

This section presents an overview of an algorithm for inserting elements into a sorted array and preserving the order of elements. This algorithm guarantees data consistency using the versioning technique.

First, we describe the layout of our sorted array. Figure 11-2 and Listing 11-6 show that there are two arrays of elements and two size fields. Additionally, one `current` field stores information about which array and size variable is currently used.

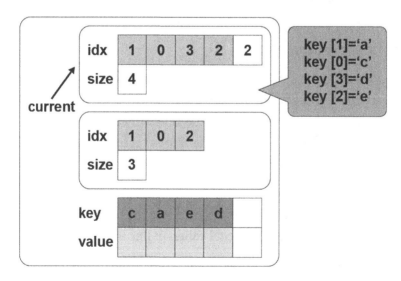

Figure 11-2. *Sorted array layout*

Listing 11-6. Sorted array layout

```
41  template <typename Value, uint64_t slots>
42  struct entries_t {
43    Value entries[slots];
44    size_t size;
45  };
46
47  template <typename Value, uint64_t slots>
48  class array {
49  public:
50    void insert(pmem::obj::pool_base &pop, const Value &);
51    void insert_element(pmem::obj::pool_base &pop, const Value&);
52
53    entries_t<Value, slots> v[2];
54    uint32_t current;
55  };
```

- Lines 41-45: We define the helper structure, which consists of an array of indexes and a size.

- Line 53: We define two elements array of entries_t structures. entries_t holds an array of elements (entries array) and the number of elements in the node as the size variable.

- Line 54: This variable determines which entries_t structure from line 53 is used. It can be only 0 or 1. Figure 11-2 shows the situation where the current is equal to 0 and points to the first element of the v array.

To understand why we need two versions of the entries_t structure and a current field, Figure 11-3 shows how the insert operation works, and the corresponding pseudocode appears in Listing 11-7.

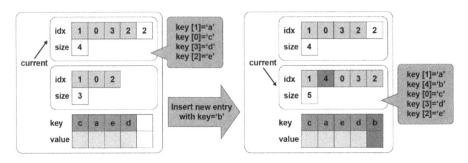

Figure 11-3. *Overview of a sorted tree insert operation*

Listing 11-7. Pseudocode of a sorted tree insert operation

```
57  template <typename Value, uint64_t slots>
58  void array<Value, slots>::insert_element(pmem::obj::pool_base &pop,
59                      const Value &entry) {
60    auto &working_copy = v[1 - current];
61    auto &consistent_copy = v[current];
62
63    auto consistent_insert_position = std::lower_bound(
64     std::begin(consistent_copy.entries),
65     std::begin(consistent_copy.entries) +
66              consistent_copy.size, entry);
67    auto working_insert_position =
68        std::begin(working_copy.entries) +
          std::distance(std::begin(consistent_copy.entries),
69        consistent_insert_position);
70
71      std::copy(std::begin(consistent_copy.entries),
72              consistent_insert_position,
73              std::begin(working_copy.entries));
74
75      *working_insert_position = entry;
76
77      std::copy(consistent_insert_position,
78              std::begin(consistent_copy.entries) +
                  consistent_copy.size,
79              working_insert_position + 1);
```

```
80
81          working_copy.size = consistent_copy.size + 1;
82  }
83
84  template <typename V, uint64_t s>
85  void array<V,s>::insert(pmem::obj::pool_base &pop,
86                                      const Value &entry){
87    insert_element(pop, entry);
88    pop.persist(&(v[1 - current]), sizeof(entries_t<Value, slots>));
89
90    current = 1 - current;
91    pop.persist(&current, sizeof(current));
92  }
```

- Lines 60-61: We define references to the current version of entries array and to the working version.

- Line 63: We find the position in the current array where an entry should be inserted.

- Line 67: We create iterator to the working array.

- Line 71: We copy part of the current array to the working array (range from beginning of the current array to the place where a new element should be inserted).

- Line 75: We insert an entry to the working array.

- Line 77: We copy remaining elements from the current array to the working array after the element we just inserted.

- Line 81: We update the size of the working array to the size of the current array plus one, for the element inserted.

- Lines 87-88: We insert an element and persist the entire v[1-current] element.

- Lines 90-91: We update the current value and save it.

Let's analyze whether this approach guarantees data consistency. In the first step, we copy elements from the original array to a currently unused one, insert the new element, and persist it to make sure data goes to the persistence domain. The persist call also ensures that the next operation (updating the current value) is not reordered before any of the previous stores. Because of this, any interruption before or after issuing the instruction to update the current field would not corrupt data because the current variable always points to a valid version.

The memory overhead of using versioning for the insert operation is equal to a size of the entries array and the current field. In terms of time overhead, we issued only two persist operations.

Summary

This chapter shows how to design data structures for persistent memory, considering its characteristics and capabilities. We discuss fragmentation and why it is problematic in the case of persistent memory. We also present a few different methods of guaranteeing data consistency; using transactions is the simplest and least error-prone method. Other approaches, such as copy-on-write or versioning, can perform better, but they are significantly more difficult to implement correctly.

CHAPTER 12

Debugging Persistent Memory Applications

Persistent memory programming introduces new opportunities that allow developers to directly persist data structures without serialization and to access them in place without involving classic block I/O. As a result, you can merge your data models and avoid the classic split between data in memory – which is volatile, fast, and byte addressable – with data on traditional storage devices, which is non-volatile but slower.

Persistent memory programming also brings challenges. Recall our discussion about power-fail protected persistence domains in Chapter 2: When a process or system crashes on an Asynchronous DRAM Refresh (ADR)-enabled platform, data residing in the CPU caches that has not yet been flushed, is lost. This is not a problem with volatile memory because all the memory hierarchy is volatile. With persistent memory, however, a crash can cause permanent data corruption. How often must you flush data? Flushing too frequently yields suboptimal performance, and not flushing often enough leaves the potential for data loss or corruption.

Chapter 11 described several approaches to designing data structures and using methods such as copy-on-write, versioning, and transactions to maintain data integrity. Many libraries within the Persistent Memory Development Kit (PMDK) provide transactional updates of data structures and variables. These libraries provide optimal CPU cache flushing, when required by the platform, at precisely the right time, so you can program without concern about the hardware intricacies.

This programming paradigm introduces new dimensions related to errors and performance issues that programmers need to be aware of. The PMDK libraries reduce errors in persistent memory programming, but they cannot eliminate them. This chapter

© The Author(s) 2020
S. Scargall, *Programming Persistent Memory*, https://doi.org/10.1007/978-1-4842-4932-1_12

describes common persistent memory programming issues and pitfalls and how to correct them using the tools available. The first half of this chapter introduces the tools. The second half presents several erroneous programming scenarios and describes how to use the tools to correct the mistakes before releasing your code into production.

pmemcheck for Valgrind

pmemcheck is a Valgrind (http://www.valgrind.org/) tool developed by Intel. It is very similar to memcheck, which is the default tool in Valgrind to discover memory-related bugs but adapted for persistent memory. Valgrind is an instrumentation framework for building dynamic analysis tools. Some Valgrind tools can automatically detect many memory management and threading bugs and profile your programs in detail. You can also use Valgrind to build new tools.

To run pmemcheck, you need a modified version of Valgrind supporting the new CLFLUSHOPT and CLWB flushing instructions. The persistent memory version of Valgrind includes the pmemcheck tool and is available from https://github.com/pmem/valgrind. Refer to the README.md within the GitHub project for installation instructions.

All the libraries in PMDK are already instrumented with pmemcheck. If you use PMDK for persistent memory programming, you will be able to easily check your code with pmemcheck without any code modification.

Before we discuss the pmemcheck details, the following two sections demonstrate how it identifies errors in an out-of-bounds and a memory leak example.

Stack Overflow Example

An out-of-bounds scenario is a stack/buffer overflow bug, where data is written or read beyond the capacity of the stack or array. Consider the small code snippet in Listing 12-1.

Listing 12-1. stackoverflow.c: Example of an out-of-bound bug

```
32  #include <stdlib.h>
33
34  int main() {
35          int *stack = malloc(100 * sizeof(int));
36          stack[100] = 1234;
```

```
37          free(stack);
38      return 0;
39  }
```

In line 36, we are incorrectly assigning the value 1234 to the position 100, which is outside the array range of 0-99. If we compile and run this code, it may not fail. This is because, even if we only allocated 400 bytes (100 integers) for our array, the operating system provides a whole memory page, typically 4KiB. Executing the binary under Valgrind reports an issue, shown in Listing 12-2.

Listing 12-2. Running Valgrind with code Listing 12-1

```
$ valgrind ./stackoverflow
==4188== Memcheck, a memory error detector
...
==4188== Invalid write of size 4
==4188==    at 0x400556: main (stackoverflow.c:36)
==4188==  Address 0x51f91d0 is 0 bytes after a block of size 400 alloc'd
==4188==    at 0x4C2EB37: malloc (vg_replace_malloc.c:299)
==4188==    by 0x400547: main (stackoverflow.c:35)
...
==4188== ERROR SUMMARY: 1 errors from 1 contexts (suppressed: 0 from 0)
```

Because Valgrind can produce long reports, we show only the relevant *"Invalid write"* error part of the report. When compiling code with symbol information (gcc -g), it is easy to see the exact place in the code where the error is detected. In this case, Valgrind highlights line 36 of the stackoverflow.c file. With the issue identified in the code, we know where to fix it.

Memory Leak Example

Memory leaks are another common issue. Consider the code in Listing 12-3.

Listing 12-3. leak.c: Example of a memory leak

```
32  #include <stdlib.h>
33
34  void func(void) {
```

```
35        int *stack = malloc(100 * sizeof(int));
36  }
37
38  int main(void) {
39        func();
40        return 0;
41  }
```

The memory allocation is moved to the function func(). A memory leak occurs because the pointer to the newly allocated memory is a local variable on line 35, which is lost when the function returns. Executing this program under Valgrind shows the results in Listing 12-4.

Listing 12-4. Running Valgrind with code Listing 12-3

```
$ valgrind --leak-check=yes ./leak
==4413== Memcheck, a memory error detector
...
==4413== 400 bytes in 1 blocks are definitely lost in loss record 1 of 1
==4413==    at 0x4C2EB37: malloc (vg_replace_malloc.c:299)
==4413==    by 0x4004F7: func (leak.c:35)
==4413==    by 0x400507: main (leak.c:39)
==4413==
==4413== LEAK SUMMARY:
...
==4413== ERROR SUMMARY: 1 errors from 1 contexts (suppressed: 0 from 0)
```

Valgrind shows a loss of 400 bytes of memory allocated at leak.c:35. To learn more, please visit the official Valgrind documentation (http://www.valgrind.org/docs/manual/index.html).

Intel Inspector – Persistence Inspector

Intel Inspector – Persistence Inspector is a runtime tool that developers use to detect programming errors in persistent memory programs. In addition to cache flush misses, this tool detects

- Redundant cache flushes and memory fences

- Out-of-order persistent memory stores

- Incorrect undo logging for the PMDK

Persistence Inspector is included as part of Intel Inspector, an easy-to-use memory and threading error debugger for C, C++, and Fortran that works with both Windows and Linux operating systems. It has an intuitive graphical and command-line interfaces, and it can be integrated with Microsoft Visual Studio. Intel Inspector is available as part of Intel Parallel Studio XE (`https://software.intel.com/en-us/parallel-studio-xe`) and Intel System Studio (`https://software.intel.com/en-us/system-studio`).

This section describes how the Intel Inspector tool works with the same out-of-bounds and memory leak examples from Listings 12-1 and 12-3.

Stack Overflow Example

The Listing 12-5 example demonstrates how to use the command-line interface to perform the analysis and collect the data and then switches to the GUI to examine the results in detail. To collect the data, we use the `inspxe-cl` utility with the `-c=mi2` collection option for detecting memory problems.

Listing 12-5. Running Intel Inspector with code Listing 12-1

```
$ inspxe-cl -c=mi2 -- ./stackoverflow

1 new problem(s) found
    1 Invalid memory access problem(s) detected
```

Intel Inspector creates a new directory with the data and analysis results, and prints a summary of findings to the terminal. For the stackoverflow app, it detected one invalid memory access.

After launching the GUI using `inspxe-gui`, we open the results collection through the *File* ➤ *Open* ➤ *Result* menu and navigate to the directory created by `inspxe-cli`. The directory will be named `r000mi2` if it is the first run. Within the directory is a file named `r000mi2.inspxe`. Once opened and processed, the GUI presents the data shown in Figure 12-1.

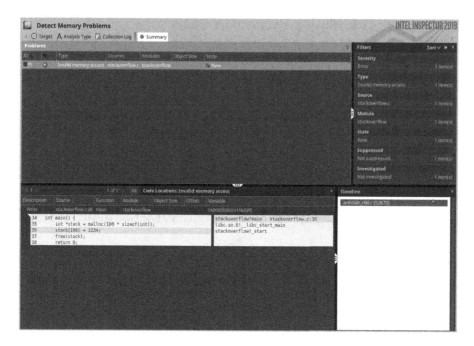

Figure 12-1. *GUI of Intel Inspector showing results for Listing 12-1*

The GUI defaults to the *Summary* tab to provide an overview of the analysis. Since we compiled the program with symbols, the *Code Locations* panel at the bottom shows the exact place in the code where the problem was detected. Intel Inspector identified the same error on line 36 that Valgrind found.

If Intel Inspector detects multiple problems within the program, those issues are listed in the *Problems* section in the upper left area of the window. You can select each problem and see the information relating to it in the other sections of the window.

Memory Leak Example

The Listing 12-6 example runs Intel Inspector using the `leak.c` code from Listing 12-2 and uses the same arguments from the `stackoverflow` program to detect memory issues.

Listing 12-6. Running Intel Inspector with code Listing 12-2

```
$ inspxe-cl -c=mi2 -- ./leak

1 new problem(s) found
    1 Memory leak problem(s) detected
```

The Intel Inspector output is shown in Figure 12-2 and explains that a memory leak problem was detected. When we open the `r001mi2/r001mi2.inspxe` result file in the GUI, we get something similar to what is shown in the lower left section of Figure 12-2.

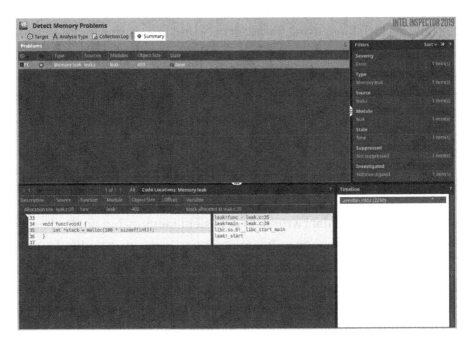

Figure 12-2. *GUI of Intel Inspector showing results for Listing 12-2*

The information related to the leaked object is shown above the code listing:

- Allocation site (source, function name, and module)

- Object size (400 bytes)

- The variable name that caused the leak

The right side of the *Code* panel shows the call stack that led to the bug (call stacks are read from bottom to top). We see the call to `func()` in the `main()` function on line 39 (`leak.c:39`), then the memory allocation occurs within `func()` on line 35 (`leak.c:35`).

The Intel Inspector offers much more than what we presented here. To learn more, please visit the documentation (`https://software.intel.com/en-us/intel-inspector-support/documentation`).

Common Persistent Memory Programming Problems

This section reviews several coding and performance problems you are likely to encounter, how to catch them using the pmemcheck and Intel Inspector tools, and how to resolve the issues.

The tools we use highlight deliberately added issues in our code that can cause bugs, data corruption, or other problems. For pmemcheck, we show how to bypass data sections that should not be checked by the tool and use macros to assist the tool in better understanding our intent.

Nonpersistent Stores

Nonpersistent stores refer to data written to persistent memory but not flushed explicitly. It is understood that if the program writes to persistent memory, it wishes for those writes to be persistent. If the program ends without explicitly flushing writes, there is an open possibility for data corruption. When a program exits gracefully, all the pending writes in the CPU caches are flushed automatically. However, if the program were to crash unexpectedly, writes still residing in the CPU caches could be lost.

Consider the code in Listing 12-7 that writes data to a persistent memory device mounted to /mnt/pmem without flushing the data.

Listing 12-7. Example of writing to persistent memory without flushing

```
32  #include <stdio.h>
33  #include <sys/mman.h>
34  #include <fcntl.h>
35
36  int main(int argc, char *argv[]) {
37      int fd, *data;
38      fd = open("/mnt/pmem/file", O_CREAT|O_RDWR, 0666);
39      posix_fallocate(fd, 0, sizeof(int));
40      data = (int *) mmap(NULL, sizeof(int), PROT_READ |
41                      PROT_WRITE, MAP_SHARED_VALIDATE |
42                      MAP_SYNC, fd, 0);
```

```
43      *data = 1234;
44      munmap(data, sizeof(int));
45      return 0;
46  }
```

- Line 38: We open /mnt/pmem/file.

- Line 39: We make sure there is enough space in the file to allocate an integer by calling posix_fallocate().

- Line 40: We memory map /mnt/pmem/file.

- Line 43: We write 1234 to the memory.

- Line 44: We unmap the memory.

If we run pmemcheck with Listing 12-7, we will not get any useful information because pmemcheck has no way to know which memory addresses are persistent and which ones are volatile. This may change in future versions. To run pmemcheck, we pass --tool=pmemcheck argument to valgrind as shown in Listing 12-8. The result shows no issues were detected.

Listing 12-8. Running pmemcheck with code Listing 12-7

```
$ valgrind --tool=pmemcheck ./listing_12-7
==116951== pmemcheck-1.0, a simple persistent store checker
==116951== Copyright (c) 2014-2016, Intel Corporation
==116951== Using Valgrind-3.14.0 and LibVEX; rerun with -h for copyright
            info
==116951== Command: ./listing_12-9
==116951==
==116951==
==116951== Number of stores not made persistent: 0
==116951== ERROR SUMMARY: 0 errors
```

We can inform pmemcheck which memory regions are persistent using a VALGRIND_PMC_REGISTER_PMEM_MAPPING macro shown on line 52 in Listing 12-9. We must include the valgrind/pmemcheck.h header for pmemcheck, line 36, which defines the VALGRIND_PMC_REGISTER_PMEM_MAPPING macro and others.

Listing 12-9. Example of writing to persistent memory using Valgrind macros without flushing

```
33   #include <stdio.h>
34   #include <sys/mman.h>
35   #include <fcntl.h>
36   #include <valgrind/pmemcheck.h>
37
38   int main(int argc, char *argv[]) {
39       int fd, *data;
40
41       // open the file and allocate enough space for an
42       // integer
43       fd = open("/mnt/pmem/file", O_CREAT|O_RDWR, 0666);
44       posix_fallocate(fd, 0, sizeof(int));
45
46       // memory map the file and register the mapped
47       // memory with VALGRIND
48       data = (int *) mmap(NULL, sizeof(int),
49               PROT_READ|PROT_WRITE,
50               MAP_SHARED_VALIDATE | MAP_SYNC,
51               fd, 0);
52       VALGRIND_PMC_REGISTER_PMEM_MAPPING(data,
53                                   sizeof(int));
54
55       // write to pmem
56       *data = 1234;
57
58       // unmap the memory and un-register it with
59       // VALGRIND
60       munmap(data, sizeof(int));
61       VALGRIND_PMC_REMOVE_PMEM_MAPPING(data,
62                                   sizeof(int));
63       return 0;
64   }
```

We remove persistent memory mapping identification from pmemcheck using the
VALGRIND_PMC_REMOVE_PMEM_MAPPING macro. As mentioned earlier, this is useful when
you want to exclude parts of persistent memory from the analysis. Listing 12-10 shows
executing pmemcheck with the modified code in Listing 12-9, which now reports a
problem.

Listing 12-10. Running pmemcheck with code Listing 12-9

```
$ valgrind --tool=pmemcheck ./listing_12-9
==8904== pmemcheck-1.0, a simple persistent store checker
...
==8904== Number of stores not made persistent: 1
==8904== Stores not made persistent properly:
==8904== [0]    at 0x4008B4: main (listing_12-9.c:56)
==8904==          Address: 0x4027000    size: 4 state: DIRTY
==8904== Total memory not made persistent: 4
==8904== ERROR SUMMARY: 1 errors
```

See that pmemcheck detected that data is not being flushed after a write in
listing_12-9.c, line 56. To fix this, we create a new flush() function, accepting an
address and size, to flush all the CPU cache lines storing any part of the data using the
CLFLUSH machine instruction (__mm_clflush()). Listing 12-11 shows the modified
code.

Listing 12-11. Example of writing to persistent memory using Valgrind with
flushing

```
33  #include <emmintrin.h>
34  #include <stdint.h>
35  #include <stdio.h>
36  #include <sys/mman.h>
37  #include <fcntl.h>
38  #include <valgrind/pmemcheck.h>
39
```

```
40   // flushing from user space
41   void flush(const void *addr, size_t len) {
42       uintptr_t flush_align = 64, uptr;
43       for (uptr = (uintptr_t)addr & ~(flush_align - 1);
44                uptr < (uintptr_t)addr + len;
45                uptr += flush_align)
46           _mm_clflush((char *)uptr);
47   }
48
49   int main(int argc, char *argv[]) {
50       int fd, *data;
51
52       // open the file and allocate space for one
53       // integer
54       fd = open("/mnt/pmem/file", O_CREAT|O_RDWR, 0666);
55       posix_fallocate(fd, 0, sizeof(int));
56
57       // map the file and register it with VALGRIND
58       data = (int *)mmap(NULL, sizeof(int),
59               PROT_READ | PROT_WRITE,
60               MAP_SHARED_VALIDATE | MAP_SYNC, fd, 0);
61       VALGRIND_PMC_REGISTER_PMEM_MAPPING(data,
62                                           sizeof(int));
63
64       // write and flush
65       *data = 1234;
66       flush((void *)data, sizeof(int));
67
68       // unmap and un-register
69       munmap(data, sizeof(int));
70       VALGRIND_PMC_REMOVE_PMEM_MAPPING(data,
71                                         sizeof(int));
72       return 0;
73   }
```

Running the modified code through pmemcheck reports no issues, as shown in Listing 12-12.

Listing 12-12. Running pmemcheck with code Listing 12-11

```
$ valgrind --tool=pmemcheck ./listing_12-11
==9710== pmemcheck-1.0, a simple persistent store checker
...
==9710== Number of stores not made persistent: 0
==9710== ERROR SUMMARY: 0 errors
```

Because Intel Inspector – Persistence Inspector does not consider an unflushed write a problem unless there is a write dependency with other variables, we need to show a more complex example than writing a single variable in Listing 12-7. You need to understand how programs writing to persistent memory are designed to know which parts of the data written to the persistent media are valid and which parts are not. Remember that recent writes may still be sitting on the CPU caches if they are not explicitly flushed.

Transactions solve the problem of half-written data by using logs to either roll back or apply uncommitted changes; thus, programs reading the data back can be assured that everything written is valid. In the absence of transactions, it is impossible to know whether or not the data written on persistent memory is valid, especially if the program crashes.

A writer can inform a reader that data is properly written in one of two ways, either by setting a "valid" flag or by using a watermark variable with the address (or the index, in the case of an array) of the last valid written memory position.

Listing 12-13 shows pseudocode for how the "valid" flag approach could be implemented.

Listing 12-13. Pseudocode showcasing write dependency of var1 with var1_valid

```
1  writer() {
2          var1 = "This is a persistent Hello World
3                  written to persistent memory!";
4          flush (var1);
5          var1_valid = True;
6          flush (var1_valid);
7  }
8
```

```
 9  reader() {
10          if (var1_valid == True) {
11                  print (var1);
12          }
14  }
```

The reader() will read the data in var1 if the var1_valid flag is set to True (line 10), and var1_valid can only be True if var1 has been flushed (lines 4 and 5).

We can now modify the code from Listing 12-7 to introduce this "valid" flag. In Listing 12-14, we separate the code into writer and reader programs and map two integers instead of one (to accommodate for the flag). Listing 12-15 shows the *reading* to persistent memory example.

Listing 12-14. Example of writing to persistent memory with a write dependency; the code does not flush

```
33  #include <stdio.h>
34  #include <sys/mman.h>
35  #include <fcntl.h>
36  #include <string.h>
37
38  int main(int argc, char *argv[]) {
39      int fd, *ptr, *data, *flag;
40
41      fd = open("/mnt/pmem/file", O_CREAT|O_RDWR, 0666);
42      posix_fallocate(fd, 0, sizeof(int)*2);
43
44      ptr = (int *) mmap(NULL, sizeof(int)*2,
45                      PROT_READ | PROT_WRITE,
46                      MAP_SHARED_VALIDATE | MAP_SYNC,
47                      fd, 0);
48
49      data = &(ptr[1]);
50      flag = &(ptr[0]);
51      *data = 1234;
52      *flag = 1;
53
```

```
54    munmap(ptr, 2 * sizeof(int));
55    return 0;
56 }
```

Listing 12-15. Example of reading from persistent memory with a write
dependency

```
33 #include <stdio.h>
34 #include <sys/mman.h>
35 #include <fcntl.h>
36
37 int main(int argc, char *argv[]) {
38     int fd, *ptr, *data, *flag;
39
40     fd = open("/mnt/pmem/file", O_CREAT|O_RDWR, 0666);
41     posix_fallocate(fd, 0, 2 * sizeof(int));
42
43     ptr = (int *) mmap(NULL, 2 * sizeof(int),
44                        PROT_READ | PROT_WRITE,
45                        MAP_SHARED_VALIDATE | MAP_SYNC,
46                        fd, 0);
47
48     data = &(ptr[1]);
49     flag = &(ptr[0]);
50     if (*flag == 1)
51         printf("data = %d\n", *data);
52
53     munmap(ptr, 2 * sizeof(int));
54     return 0;
55 }
```

Checking our code with Persistence Inspector is done in three steps.

Step 1: We must run the before-unfortunate-event phase analysis (see Listing 12-16),
which corresponds to the writer code in Listing 12-14.

Listing 12-16. Running Intel Inspector – Persistence Inspector with code Listing 12-14 for before-unfortunate-event phase analysis

```
$ pmeminsp cb -pmem-file /mnt/pmem/file -- ./listing_12-14
++ Analysis starts

++ Analysis completes
++ Data is stored in folder "/data/.pmeminspdata/data/listing_12-14"
```

The parameter `cb` is an abbreviation of *check-before-unfortunate-event*, which specifies the type of analysis. We must also pass the persistent memory file that will be used by the application so that Persistence Inspector knows which memory accesses correspond to persistent memory. By default, the output of the analysis is stored in a local directory under the `.pmeminspdata` directory. (You can also specify a custom directory; run `pmeminsp -help` for information on the available options.)

Step 2: We run the *after-unfortunate-event* phase analysis (see Listing 12-17). This corresponds to the code that will read the data after an unfortunate event happens, such as a process crash.

Listing 12-17. Running Intel Inspector – Persistence Inspector with code Listing 12-15 for after-unfortunate-event phase analysis

```
$ pmeminsp ca -pmem-file /mnt/pmem/file -- ./listing_12-15
++ Analysis starts

data = 1234

++ Analysis completes
++ Data is stored in folder "/data/.pmeminspdata/data/listing_12-15"
```

The parameter `ca` is an abbreviation of *check-after-unfortunate-event*. Again, the output of the analysis is stored in `.pmeminspdata` within the current working directory.

Step 3: We generate the final report. For this, we pass the option `rp` (abbreviation for `report`) along with the name of both programs, as shown in Listing 12-18.

Listing 12-18. Generating a final report with Intel Inspector – Persistence Inspector from the analysis done in Listings 12-16 and 12-17

```
$ pmeminsp rp -- listing_12-16 listing_12-17
#================================================================
# Diagnostic # 1: Missing cache flush
#-------------------
  The first memory store
    of size 4 at address 0x7F9C68893004 (offset 0x4 in /mnt/pmem/file)
    in /data/listing_12-16!main at listing_12-16.c:51 - 0x67D
    in /lib64/libc.so.6!__libc_start_main at <unknown_file>:<unknown_
    line> - 0x223D3
    in /data/listing_12-16!_start at <unknown_file>:<unknown_line> - 0x534

  is not flushed before

  the second memory store
    of size 4 at address 0x7F9C68893000 (offset 0x0 in /mnt/pmem/file)
    in /data/listing_12-16!main at listing_12-16.c:52 - 0x687
    in /lib64/libc.so.6!__libc_start_main at <unknown_file>:<unknown_
    line> - 0x223D3
    in /data/listing_12-16!_start at <unknown_file>:<unknown_line> - 0x534

  while

  memory load from the location of the first store
    in /data/listing_12-17!main at listing_12-17.c:51 - 0x6C8

  depends on

  memory load from the location of the second store
    in /data/listing_12-17!main at listing_12-17.c:50 - 0x6BD

#================================================================
# Diagnostic # 2: Missing cache flush
#-------------------
  Memory store
    of size 4 at address 0x7F9C68893000 (offset 0x0 in /mnt/pmem/file)
    in /data/listing_12-16!main at listing_12-16.c:52 - 0x687
```

```
    in /lib64/libc.so.6!__libc_start_main at <unknown_file>:<unknown_
    line> - 0x223D3
    in /data/listing_12-16!_start at <unknown_file>:<unknown_line> - 0x534
```

  is not flushed before

memory is unmapped

```
    in /data/listing_12-16!main at listing_12-16.c:54 - 0x699
    in /lib64/libc.so.6!__libc_start_main at <unknown_file>:<unknown_
    line> - 0x223D3
    in /data/listing_12-16!_start at <unknown_file>:<unknown_line> - 0x534
```

Analysis complete. 2 diagnostic(s) reported.

The output is very verbose, but it is easy to follow. We get two missing cache flushes (diagnostics 1 and 2) corresponding to lines 51 and 52 of listing_12-16.c. We do these writes to the locations in the mapped persistent memory pointed by variables flag and data. The first diagnostic says that the first memory store is not flushed before the second store, while, at the same time, there is a load dependency of the first store to the second. This is exactly what we intended.

The second diagnostic says that the second store (to the flag) itself is never actually flushed before ending. Even if we flush the first store correctly before we write the flag, we must still flush the flag to make sure the dependency works.

To open the results in the Intel Inspector GUI, you can use the -insp option when generating the report, for example:

```
$ pmeminsp rp -insp -- listing_12-16 listing_12-17
```

This generates a directory called r000pmem inside the analysis directory (.pmeminspdata by default). Launch the GUI running inspxe-gui and open the result file by going to *File ➤ Open ➤ Result* and selecting the file r000pmem/r000pmem.inspxe. You should see something similar to what is shown in Figure 12-3.

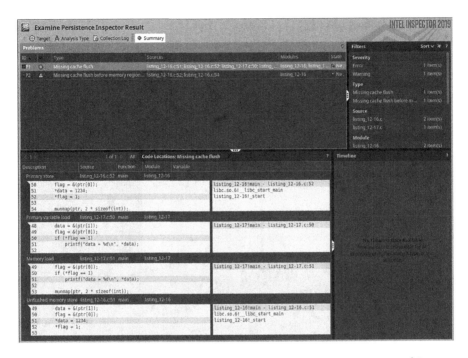

Figure 12-3. *GUI of Intel Inspector showing results for Listing 12-18 (diagnostic 1)*

The GUI shows the same information as the command-line analysis but in a more readable way by highlighting the errors directly on our source code. As Figure 12-3 shows, the modification of the flag is called "primary store."

In Figure 12-4, the second diagnosis is selected in the Problems pane, showing the missing flush for the flag itself.

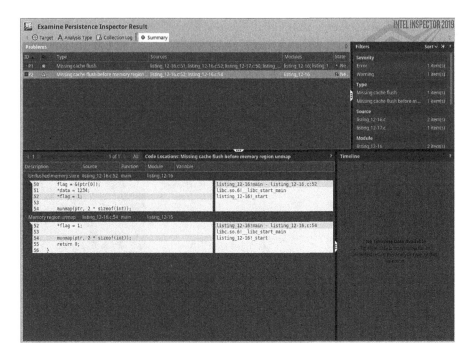

Figure 12-4. *GUI of Intel Inspector showing results for Listing 12-20 (diagnostic #2)*

To conclude this section, we fix the code and rerun the analysis with Persistence Inspector. The code in Listing 12-19 adds the necessary flushes to Listing 12-14.

Listing 12-19. Example of writing to persistent memory with a write dependency. The code flushes both writes

```
33  #include <emmintrin.h>
34  #include <stdint.h>
35  #include <stdio.h>
36  #include <sys/mman.h>
37  #include <fcntl.h>
38  #include <string.h>
39
40  void flush(const void *addr, size_t len) {
41      uintptr_t flush_align = 64, uptr;
42      for (uptr = (uintptr_t)addr & ~(flush_align - 1);
43              uptr < (uintptr_t)addr + len;
44              uptr += flush_align)
```

```
45          _mm_clflush((char *)uptr);
46  }
47
48  int main(int argc, char *argv[]) {
49      int fd, *ptr, *data, *flag;
50
51      fd = open("/mnt/pmem/file", O_CREAT|O_RDWR, 0666);
52      posix_fallocate(fd, 0, sizeof(int) * 2);
53
54      ptr = (int *) mmap(NULL, sizeof(int) * 2,
55                         PROT_READ | PROT_WRITE,
56                         MAP_SHARED_VALIDATE | MAP_SYNC,
57                         fd, 0);
58
59      data = &(ptr[1]);
60      flag = &(ptr[0]);
61      *data = 1234;
62      flush((void *) data, sizeof(int));
63      *flag = 1;
64      flush((void *) flag, sizeof(int));
65
66      munmap(ptr, 2 * sizeof(int));
67      return 0;
68  }
```

Listing 12-20 executes Persistence Inspector against the modified code from
Listing 12-19, then the reader code from Listing 12-15, and finally running the report,
which says that no problems were detected.

Listing 12-20. Running full analysis with Intel Inspector – Persistence Inspector
with code Listings 12-19 and 12-15

```
$ pmeminsp cb -pmem-file /mnt/pmem/file -- ./listing_12-19
++ Analysis starts

++ Analysis completes
++ Data is stored in folder "/data/.pmeminspdata/data/listing_12-19"
```

```
$ pmeminsp ca -pmem-file /mnt/pmem/file -- ./listing_12-15
++ Analysis starts

data = 1234

++ Analysis completes
++ Data is stored in folder "/data/.pmeminspdata/data/listing_12-15"

$ pmeminsp rp -- listing_12-19 listing_12-15
Analysis complete. No problems detected.
```

Stores Not Added into a Transaction

When working within a transaction block, it is assumed that all the modified persistent memory addresses were added to it at the beginning, which also implies that their previous values are copied to an undo log. This allows the transaction to implicitly flush added memory addresses at the end of the block or roll back to the old values in the event of an unexpected failure. A modification within a transaction to an address that is not added to the transaction is a bug that you must be aware of.

Consider the code in Listing 12-21 that uses the libpmemobj library from PMDK. It shows an example of writing within a transaction using a memory address that is not explicitly tracked by the transaction.

Listing 12-21. Example of writing within a transaction with a memory address not added to the transaction

```
33   #include <libpmemobj.h>
34
35   struct my_root {
36       int value;
37       int is_odd;
38   };
39
40   // registering type 'my_root' in the layout
41   POBJ_LAYOUT_BEGIN(example);
42   POBJ_LAYOUT_ROOT(example, struct my_root);
43   POBJ_LAYOUT_END(example);
44
```

```
45  int main(int argc, char *argv[]) {
46      // creating the pool
47      PMEMobjpool *pop= pmemobj_create("/mnt/pmem/pool",
48                      POBJ_LAYOUT_NAME(example),
49                      (1024 * 1024 * 100), 0666);
50
51      // transation
52      TX_BEGIN(pop) {
53          TOID(struct my_root) root
54              = POBJ_ROOT(pop, struct my_root);
55
56          // adding root.value to the transaction
57          TX_ADD_FIELD(root, value);
58
59          D_RW(root)->value = 4;
60          D_RW(root)->is_odd = D_RO(root)->value % 2;
61      } TX_END
62
63      return 0;
64  }
```

Note   For a refresh on the definitions of a layout, root object, or macros used in Listing 12-21, see Chapter 7 where we introduce libpmemobj.

In lines 35-38, we create a my_root data structure, which has two integer members: value and is_odd. These integers are modified inside a transaction (lines 52-61), setting value=4 and is_odd=0. On line 57, we are only adding the value variable to the transaction, leaving is_odd out. Given that persistent memory is not natively supported in C, there is no way for the compiler to warn you about this. The compiler cannot distinguish between pointers to volatile memory vs. those to persistent memory.

Listing 12-22 shows the response from running the code through pmemcheck.

Listing 12-22. Running pmemcheck with code Listing 12-21

```
$ valgrind --tool=pmemcheck ./listing_12-21
==48660== pmemcheck-1.0, a simple persistent store checker
==48660== Copyright (c) 2014-2016, Intel Corporation
==48660== Using Valgrind-3.14.0 and LibVEX; rerun with -h for copyright info
==48660== Command: ./listing_12-21
==48660==
==48660==
==48660== Number of stores not made persistent: 1
==48660== Stores not made persistent properly:
==48660== [0]    at 0x400C2D: main (listing_12-25.c:60)
==48660==        Address: 0x7dc0554     size: 4 state: DIRTY
==48660== Total memory not made persistent: 4
==48660==
==48660== Number of stores made without adding to transaction: 1
==48660== Stores made without adding to transactions:
==48660== [0]    at 0x400C2D: main (listing_12-25.c:60)
==48660==        Address: 0x7dc0554     size: 4
==48660== ERROR SUMMARY: 2 errors
```

Although they are both related to the same root cause, pmemcheck identified two issues. One is the error we expected; that is, we have a store inside a transaction that was not added to it. The other error says that we are not flushing the store. Since transactional stores are flushed automatically when the program exits the transaction, finding two errors per store to a location not included within a transaction should be common in pmemcheck.

Persistence Inspector has a more user-friendly output, as shown in Listing 12-23.

Listing 12-23. Generating a report with Intel Inspector – Persistence Inspector for code Listing 12-21

```
$ pmeminsp cb -pmem-file /mnt/pmem/pool -- ./listing_12-21
++ Analysis starts

++ Analysis completes
++ Data is stored in folder "/data/.pmeminspdata/data/listing_12-21"
$
```

```
$ pmeminsp rp -- ./listing_12-21
#==============================================================
# Diagnostic # 1: Store without undo log
#-------------------
  Memory store
    of size 4 at address Ox7FAA84DC0554 (offset Ox3C0554 in /mnt/pmem/pool)
    in /data/listing_12-21!main at listing_12-21.c:60 - OxC2D
    in /lib64/libc.so.6!__libc_start_main at <unknown_file>:<unknown_
    line> - Ox223D3
    in /data/listing_12-21!_start at <unknown_file>:<unknown_line> - Ox954

  is not undo logged in

  transaction
    in /data/listing_12-21!main at listing_12-21.c:52 - OxB67
    in /lib64/libc.so.6!__libc_start_main at <unknown_file>:<unknown_
    line> - Ox223D3
    in /data/listing_12-21!_start at <unknown_file>:<unknown_line> - Ox954

Analysis complete. 1 diagnostic(s) reported.
```

We do not perform an *after-unfortunate-event* phase analysis here because we are
only concerned about transactions.

We can fix the problem reported in Listing 12-23 by adding the whole root object to
the transaction using TX_ADD(root), as shown on line 53 in Listing 12-24.

Listing 12-24. Example of adding an object and writing it within a transaction

```
32   #include <libpmemobj.h>
33
34   struct my_root {
35       int value;
36       int is_odd;
37   };
38
39   POBJ_LAYOUT_BEGIN(example);
40   POBJ_LAYOUT_ROOT(example, struct my_root);
41   POBJ_LAYOUT_END(example);
42
```

```
43  int main(int argc, char *argv[]) {
44      PMEMobjpool *pop= pmemobj_create("/mnt/pmem/pool",
45                      POBJ_LAYOUT_NAME(example),
46                      (1024 * 1024 * 100), 0666);
47
48      TX_BEGIN(pop) {
49          TOID(struct my_root) root
50              = POBJ_ROOT(pop, struct my_root);
51
52          // adding full root to the transaction
53          TX_ADD(root);
54
55          D_RW(root)->value = 4;
56          D_RW(root)->is_odd = D_RO(root)->value % 2;
57      } TX_END
58
59      return 0;
60  }
```

If we run the code through pmemcheck, as shown in Listing 12-25, no issues are reported.

Listing 12-25. Running pmemcheck with code Listing 12-24

```
$ valgrind --tool=pmemcheck ./listing_12-24
==80721== pmemcheck-1.0, a simple persistent store checker
==80721== Copyright (c) 2014-2016, Intel Corporation
==80721== Using Valgrind-3.14.0 and LibVEX; rerun with -h for copyright
          info
==80721== Command: ./listing_12-24
==80721==
==80721==
==80721== Number of stores not made persistent: 0
==80721== ERROR SUMMARY: 0 errors
```

Similarly, no issues are reported by Persistence Inspector in Listing 12-26.

Listing 12-26. Generating report with Intel Inspector – Persistence Inspector for code Listing 12-24

```
$ pmeminsp cb -pmem-file /mnt/pmem/pool -- ./listing_12-24
++ Analysis starts

++ Analysis completes
++ Data is stored in folder "/data/.pmeminspdata/data/listing_12-24"
$
$ pmeminsp rp -- ./listing_12-24
Analysis complete. No problems detected.
```

After properly adding all the memory that will be modified to the transaction, both tools report that no problems were found.

Memory Added to Two Different Transactions

In the case where one program can work with multiple transactions simultaneously, adding the same memory object to multiple transactions can potentially corrupt data. This can occur in PMDK, for example, where the library maintains a different transaction per thread. If two threads write to the same object within different transactions, after an application crash, a thread might overwrite modifications made by another thread in a different transaction. In database systems, this problem is known as *dirty reads.* Dirty reads violate the isolation requirement of the ACID (atomicity, consistency, isolation, durability) properties, as shown in Figure 12-5.

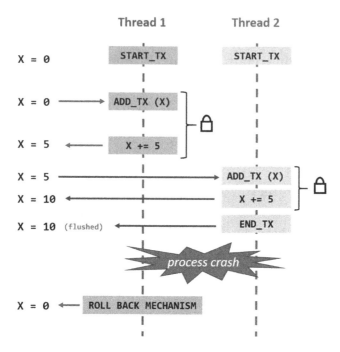

Figure 12-5. *The rollback mechanism for the unfinished transaction in Thread 1 is also overriding the changes made by Thread 2, even though the transaction for Thread 2 finishes correctly*

In Figure 12-5, time is shown in the y axis with time progressing downward. These operations occur in the following order:

- Assume X=0 when the application starts.

- A main() function creates two threads: Thread 1 and Thread 2. Both threads are intended to start their own transactions and acquire the lock to modify X.

- Since Thread 1 runs first, it acquires the lock on X first. It then adds the X variable to the transaction before incrementing X by 5. Transparent to the program, the value of X (X=0) is added to the undo log when X was added to the transaction. Since the transaction is not yet complete, the application has not yet explicitly flushed the value.

- Thread 2 starts, begins its own transaction, acquires the lock, reads the value of X (which is now 5), adds X=5 to the undo log, and increments it by 5. The transaction completes successfully, and Thread 2 flushes the CPU caches. Now, x=10.

- Unfortunately, the program crashes after Thread 2 successfully completes its transaction but before Thread 1 was able to finish its transaction and flush its value.

This scenario leaves the application with an invalid, but consistent, value of x=10. Since transactions are atomic, all changes done within them are not valid until they successfully complete.

When the application starts, it knows it must perform a recovery operation due to the previous crash and will replay the undo logs to rewind the partial update made by Thread 1. The undo log restores the value of X=0, which was correct when Thread 1 added its entry. The expected value of X should be X=5 in this situation, but the undo log puts X=0. You can probably see the huge potential for data corruption that this situation can produce.

We describe concurrency for multithreaded applications in Chapter 14. Using `libpmemobj-cpp`, the C++ language binding library to `libpmemobj`, concurrency issues are very easy to resolve because the API allows us to pass a list of locks using lambda functions when transactions are created. Chapter 8 discusses `libpmemobj-cpp` and lambda functions in more detail.

Listing 12-27 shows how you can use a single mutex to lock a whole transaction. This mutex can either be a standard mutex (`std::mutex`) if the mutex object resides in volatile memory or a pmem mutex (`pmem::obj::mutex`) if the mutex object resides in persistent memory.

Listing 12-27. Example of a libpmemobj++ transaction whose writes are both atomic – with respect to persistent memory – and isolated – in a multithreaded scenario. The mutex is passed to the transaction as a parameter

```
transaction::run (pop, [&] {
    ...
    // all writes here are atomic and thread safe
    ...
}, mutex);
```

Consider the code in Listing 12-28 that simultaneously adds the same memory region to two different transactions.

Listing 12-28. Example of two threads simultaneously adding the same persistent memory location to their respective transactions

```
33  #include <libpmemobj.h>
34  #include <pthread.h>
35
36  struct my_root {
37      int value;
38      int is_odd;
39  };
40
41  POBJ_LAYOUT_BEGIN(example);
42  POBJ_LAYOUT_ROOT(example, struct my_root);
43  POBJ_LAYOUT_END(example);
44
45  pthread_mutex_t lock;
46
47  // function to be run by extra thread
48  void *func(void *args) {
49      PMEMobjpool *pop = (PMEMobjpool *) args;
50
51      TX_BEGIN(pop) {
52          pthread_mutex_lock(&lock);
53          TOID(struct my_root) root
54              = POBJ_ROOT(pop, struct my_root);
55          TX_ADD(root);
56          D_RW(root)->value = D_RO(root)->value + 3;
57          pthread_mutex_unlock(&lock);
58      } TX_END
59  }
60
61  int main(int argc, char *argv[]) {
62      PMEMobjpool *pop= pmemobj_create("/mnt/pmem/pool",
63                      POBJ_LAYOUT_NAME(example),
64                      (1024 * 1024 * 10), 0666);
65
```

```
66      pthread_t thread;
67      pthread_mutex_init(&lock, NULL);
68
69      TX_BEGIN(pop) {
70          pthread_mutex_lock(&lock);
71          TOID(struct my_root) root
72              = POBJ_ROOT(pop, struct my_root);
73          TX_ADD(root);
74          pthread_create(&thread, NULL,
75                          func, (void *) pop);
76          D_RW(root)->value = D_RO(root)->value + 4;
77          D_RW(root)->is_odd = D_RO(root)->value % 2;
78          pthread_mutex_unlock(&lock);
79          // wait to make sure other thread finishes 1st
80          pthread_join(thread, NULL);
81      } TX_END
82
83      pthread_mutex_destroy(&lock);
84      return 0;
85  }
```

- Line 69: The main thread starts a transaction and adds the root data structure to it (line 73).

- Line 74: We create a new thread by calling pthread_create() and have it execute the func() function. This function also starts a transaction (line 51) and adds the root data structure to it (line 55).

- Both threads will simultaneously modify all or part of the same data before finishing their transactions. We force the second thread to finish first by making the main thread wait on pthread_join().

Listing 12-29 shows code execution with pmemcheck, and the result warns us that we have *overlapping regions registered in different transactions.*

Listing 12-29. Running pmemcheck with Listing 12-28

```
$ valgrind --tool=pmemcheck ./listing_12-28
==97301== pmemcheck-1.0, a simple persistent store checker
==97301== Copyright (c) 2014-2016, Intel Corporation
==97301== Using Valgrind-3.14.0 and LibVEX; rerun with -h for copyright info
==97301== Command: ./listing_12-28
==97301==
==97301==
==97301== Number of stores not made persistent: 0
==97301==
==97301== Number of overlapping regions registered in different
          transactions: 1
==97301== Overlapping regions:
==97301== [0]    at 0x4E6B0BC: pmemobj_tx_add_snapshot (in /usr/lib64/
               libpmemobj.so.1.0.0)
==97301==     by 0x4E6B5F8: pmemobj_tx_add_common.constprop.18 (in /usr/
               lib64/libpmemobj.so.1.0.0)
==97301==     by 0x4E6C62F: pmemobj_tx_add_range (in /usr/lib64/libpmemobj.
               so.1.0.0)
==97301==     by 0x400DAC: func (listing_12-28.c:55)
==97301==     by 0x4C2DDD4: start_thread (in /usr/lib64/libpthread-2.17.so)
==97301==     by 0x5180EAC: clone (in /usr/lib64/libc-2.17.so)
==97301==    Address: 0x7dc0550   size: 8    tx_id: 2
==97301==    First registered here:
==97301== [0]'   at 0x4E6B0BC: pmemobj_tx_add_snapshot (in /usr/lib64/
               libpmemobj.so.1.0.0)
==97301==     by 0x4E6B5F8: pmemobj_tx_add_common.constprop.18 (in /usr/
               lib64/libpmemobj.so.1.0.0)
==97301==     by 0x4E6C62F: pmemobj_tx_add_range (in /usr/lib64/libpmemobj.
               so.1.0.0)
==97301==     by 0x400F23: main (listing_12-28.c:73)
==97301==    Address: 0x7dc0550   size: 8    tx_id: 1
==97301== ERROR SUMMARY: 1 errors
```

Listing 12-30 shows the same code run with Persistence Inspector, which also reports *"Overlapping regions registered in different transactions"* in diagnostic 25. The first 24 diagnostic results were related to stores not added to our transactions corresponding with the locking and unlocking of our volatile mutex; these can be ignored.

Listing 12-30. Generating a report with Intel Inspector – Persistence Inspector for code Listing 12-28

```
$ pmeminsp rp -- ./listing_12-28
...
#==============================================================
# Diagnostic # 25: Overlapping regions registered in different transactions
#-------------------
  transaction
    in /data/listing_12-28!main at listing_12-28.c:69 - 0xEB6
    in /lib64/libc.so.6!__libc_start_main at <unknown_file>:<unknown_line>
    - 0x223D3
    in /data/listing_12-28!_start at <unknown_file>:<unknown_line> - 0xB44

  protects

  memory region
    in /data/listing_12-28!main at listing_12-28.c:73 - 0xF1F
    in /lib64/libc.so.6!__libc_start_main at <unknown_file>:<unknown_line>
    - 0x223D3
    in /data/listing_12-28!_start at <unknown_file>:<unknown_line> - 0xB44

  overlaps with

  memory region
    in /data/listing_12-28!func at listing_12-28.c:55 - 0xDA8
    in /lib64/libpthread.so.0!start_thread at <unknown_file>:<unknown_line>
    - 0x7DCD
    in /lib64/libc.so.6!__clone at <unknown_file>:<unknown_line> - 0xFDEAB

Analysis complete. 25 diagnostic(s) reported.
```

Memory Overwrites

When multiple modifications to the same persistent memory location occur before the location is made persistent (that is, flushed), a memory overwrite occurs. This is a potential data corruption source if a program crashes because the final value of the persistent variable can be any of the values written between the last flush and the crash. It is important to know that this may not be an issue if it is in the code by design. We recommend using volatile variables for short-lived data and only write to persistent variables when you want to persist data.

Consider the code in Listing 12-31, which writes twice to the data variable inside the main() function (lines 62 and 63) before we call flush() on line 64.

Listing 12-31. Example of persistent memory overwriting – variable data – before flushing

```
33  #include <emmintrin.h>
34  #include <stdint.h>
35  #include <stdio.h>
36  #include <sys/mman.h>
37  #include <fcntl.h>
38  #include <valgrind/pmemcheck.h>
39
40  void flush(const void *addr, size_t len) {
41      uintptr_t flush_align = 64, uptr;
42      for (uptr = (uintptr_t)addr & ~(flush_align - 1);
43              uptr < (uintptr_t)addr + len;
44              uptr += flush_align)
45          _mm_clflush((char *)uptr);
46  }
47
48  int main(int argc, char *argv[]) {
49      int fd, *data;
50
51      fd = open("/mnt/pmem/file", O_CREAT|O_RDWR, 0666);
52      posix_fallocate(fd, 0, sizeof(int));
53
```

```
54      data = (int *)mmap(NULL, sizeof(int),
55              PROT_READ | PROT_WRITE,
56              MAP_SHARED_VALIDATE | MAP_SYNC,
57              fd, 0);
58      VALGRIND_PMC_REGISTER_PMEM_MAPPING(data,
59                                          sizeof(int));
60
61      // writing twice before flushing
62      *data = 1234;
63      *data = 4321;
64      flush((void *)data, sizeof(int));
65
66      munmap(data, sizeof(int));
67      VALGRIND_PMC_REMOVE_PMEM_MAPPING(data,
68                                        sizeof(int));
69      return 0;
70  }
```

Listing 12-32 shows the report from pmemcheck with the code from Listing 12-31. To make pmemcheck look for overwrites, we must use the --mult-stores=yes option.

Listing 12-32. Running pmemcheck with Listing 12-31

```
$ valgrind --tool=pmemcheck --mult-stores=yes ./listing_12-31
==25609== pmemcheck-1.0, a simple persistent store checker
==25609== Copyright (c) 2014-2016, Intel Corporation
==25609== Using Valgrind-3.14.0 and LibVEX; rerun with -h for copyright info
==25609== Command: ./listing_12-31
==25609==
==25609==
==25609== Number of stores not made persistent: 0
==25609==
==25609== Number of overwritten stores: 1
==25609== Overwritten stores before they were made persistent:
==25609== [0]    at 0x400962: main (listing_12-31.c:62)
==25609==        Address: 0x4023000     size: 4 state: DIRTY
==25609== ERROR SUMMARY: 1 errors
```

pmemcheck reports that we have overwritten stores. We can fix this problem by either inserting a flushing instruction between both writes, if we forgot to flush, or by moving one of the stores to volatile data if that store corresponds to short-lived data.

At the time of publication, Persistence Inspector does not support checking for overwritten stores. As you have seen, Persistence Inspector does not consider a missing flush an issue unless there is a write dependency. In addition, it does not consider this a performance problem because writing to the same variable in a short time span is likely to hit the CPU caches anyway, rendering the latency differences between DRAM and persistent memory irrelevant.

Unnecessary Flushes

Flushing should be done carefully. Detecting unnecessary flushes, such as redundant ones, can help improve code performance. The code in Listing 12-33 shows a redundant call to the flush() function on line 64.

Listing 12-33. Example of redundant flushing of a persistent memory variable

```
33  #include <emmintrin.h>
34  #include <stdint.h>
35  #include <stdio.h>
36  #include <sys/mman.h>
37  #include <fcntl.h>
38  #include <valgrind/pmemcheck.h>
39
40  void flush(const void *addr, size_t len) {
41      uintptr_t flush_align = 64, uptr;
42      for (uptr = (uintptr_t)addr & ~(flush_align - 1);
43              uptr < (uintptr_t)addr + len;
44              uptr += flush_align)
45          _mm_clflush((char *)uptr);
46  }
47
48  int main(int argc, char *argv[]) {
49      int fd, *data;
50
```

```
51      fd = open("/mnt/pmem/file", O_CREAT|O_RDWR, 0666);
52      posix_fallocate(fd, 0, sizeof(int));
53
54      data = (int *)mmap(NULL, sizeof(int),
55              PROT_READ | PROT_WRITE,
56              MAP_SHARED_VALIDATE | MAP_SYNC,
57              fd, 0);
58
59      VALGRIND_PMC_REGISTER_PMEM_MAPPING(data,
60                                          sizeof(int));
61                                                       .
62      *data = 1234;
63      flush((void *)data, sizeof(int));
64      flush((void *)data, sizeof(int)); // extra flush
65
66      munmap(data, sizeof(int));
67      VALGRIND_PMC_REMOVE_PMEM_MAPPING(data,
68                                          sizeof(int));
69      return 0;
70  }
```

We can use pmemcheck to detect redundant flushes using --flush-check=yes option, as shown in Listing 12-34.

Listing 12-34. Running pmemcheck with Listing 12-33

```
$ valgrind --tool=pmemcheck --flush-check=yes ./listing_12-33
==104125== pmemcheck-1.0, a simple persistent store checker
==104125== Copyright (c) 2014-2016, Intel Corporation
==104125== Using Valgrind-3.14.0 and LibVEX; rerun with -h for copyright info
==104125== Command: ./listing_12-33
==104125==
==104125==
==104125== Number of stores not made persistent: 0
==104125==
```

```
==104125== Number of unnecessary flushes: 1
==104125== [0]    at 0x400868: flush (emmintrin.h:1459)
==104125==    by 0x400989: main (listing_12-33.c:64)
==104125==        Address: 0x4023000      size: 64
==104125== ERROR SUMMARY: 1 errors
```

To showcase Persistence Inspector, Listing 12-35 has code with a write dependency, similar to what we did for Listing 12-11 in Listing 12-19. The extra flush occurs on line 65.

Listing 12-35. Example of writing to persistent memory with a write dependency. The code does an extra flush for the flag

```
33  #include <emmintrin.h>
34  #include <stdint.h>
35  #include <stdio.h>
36  #include <sys/mman.h>
37  #include <fcntl.h>
38  #include <string.h>
39
40  void flush(const void *addr, size_t len) {
41      uintptr_t flush_align = 64, uptr;
42      for (uptr = (uintptr_t)addr & ~(flush_align - 1);
43              uptr < (uintptr_t)addr + len;
44              uptr += flush_align)
45          _mm_clflush((char *)uptr);
46  }
47
48  int main(int argc, char *argv[]) {
49      int fd, *ptr, *data, *flag;
50
51      fd = open("/mnt/pmem/file", O_CREAT|O_RDWR, 0666);
52      posix_fallocate(fd, 0, sizeof(int) * 2);
53
54      ptr = (int *) mmap(NULL, sizeof(int) * 2,
55              PROT_READ | PROT_WRITE,
56              MAP_SHARED_VALIDATE | MAP_SYNC,
57              fd, 0);
```

```
58        data = &(ptr[1]);
59        flag = &(ptr[0]);
60
61        *data = 1234;
62        flush((void *) data, sizeof(int));
63        *flag = 1;
64        flush((void *) flag, sizeof(int));
65        flush((void *) flag, sizeof(int)); // extra flush
66
67        munmap(ptr, 2 * sizeof(int));
68        return 0;
69    }
```

Listing 12-36 uses the same reader program from Listing 12-15 to show the analysis from Persistence Inspector. As before, we first collect data from the writer program, then the reader program, and finally run the report to identify any issues.

Listing 12-36. Running Intel Inspector – Persistence Inspector with Listing 12-35 (writer) and Listing 12-15 (reader)

```
$ pmeminsp cb -pmem-file /mnt/pmem/file -- ./listing_12-35
++ Analysis starts

++ Analysis completes
++ Data is stored in folder "/data/.pmeminspdata/data/listing_12-35"

$ pmeminsp ca -pmem-file /mnt/pmem/file -- ./listing_12-15
++ Analysis starts

data = 1234

++ Analysis completes
++ Data is stored in folder "/data/.pmeminspdata/data/listing_12-15"

$ pmeminsp rp -- ./listing_12-35 ./listing_12-15
#============================================================
# Diagnostic # 1: Redundant cache flush
#-------------------
  Cache flush
```

```
of size 64 at address 0x7F3220C55000 (offset 0x0 in /mnt/pmem/file)
in /data/listing_12-35!flush at listing_12-35.c:45 - 0x674
in /data/listing_12-35!main at listing_12-35.c:64 - 0x73F
in /lib64/libc.so.6!__libc_start_main at <unknown_file>:<unknown_line>
- 0x223D3
in /data/listing_12-35!_start at <unknown_file>:<unknown_line> - 0x574
```

is redundant with regard to

```
 cache flush
  of size 64 at address 0x7F3220C55000 (offset 0x0 in /mnt/pmem/file)
  in /data/listing_12-35!flush at listing_12-35.c:45 - 0x674
  in /data/listing_12-35!main at listing_12-35.c:65 - 0x750
  in /lib64/libc.so.6!__libc_start_main at <unknown_file>:<unknown_line>
  - 0x223D3
  in /data/listing_12-35!_start at <unknown_file>:<unknown_line> - 0x574
```

of

```
memory store
  of size 4 at address 0x7F3220C55000 (offset 0x0 in /mnt/pmem/file)
  in /data/listing_12-35!main at listing_12-35.c:63 - 0x72D
  in /lib64/libc.so.6!__libc_start_main at <unknown_file>:<unknown_line>
  - 0x223D3
  in /data/listing_12-35!_start at <unknown_file>:<unknown_line> - 0x574
```

The Persistence Inspector report warns about the redundant cache flush within the main() function on line 65 of the listing_12-35.c program file – "main at listing_12-35.c:65". Solving these issues is as easy as deleting all the unnecessary flushes, and the result will improve the application's performance.

Out-of-Order Writes

When developing software for persistent memory, remember that even if a cache line is not explicitly flushed, that does not mean the data is still in the CPU caches. For example, the CPU could have evicted it due to cache pressure or other reasons. Furthermore, the same way that writes that are not flushed properly may produce bugs in the event of an unexpected application crash, so do automatically evicted dirty cache lines if they violate some expected order of writes that the applications rely on.

To better understand this problem, explore how flushing works in the x86_64 and AMD64 architectures. From the user space, we can issue any of the following instructions to ensure our writes reach the persistent media:

- CLFLUSH

- CLFLUSHOPT (needs SFENCE)

- CLWB (needs SFENCE)

- Non-temporal stores (needs SFENCE)

The only instruction that ensures each flush is issued in order is CLFUSH because each CLFLUSH instruction always does an implicit fence instruction (SFENCE). The other instructions are asynchronous and can be issued in parallel and in any order. The CPU can only guarantee that all flushes issued since the previous SFENCE have completed when a new SFENCE instruction is explicitly executed. Think of SFENCE instructions as synchronization points (see Figure 12-6). For more information about these instructions, refer to the Intel software developer manuals and the AMD software developer manuals.

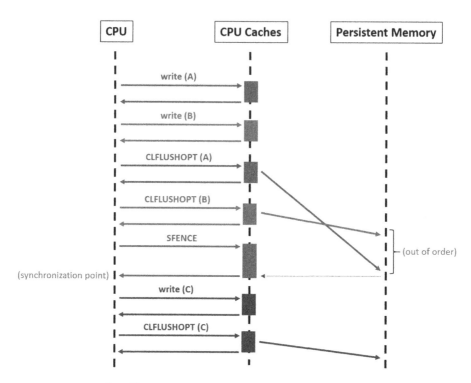

Figure 12-6. *Example of how asynchronous flushing works. The SFENCE instruction ensures a synchronization point between the writes to A and B on one side and to C on the other side*

As Figure 12-6 shows, we cannot guarantee the order with respect to how A and B would be finally written to persistent memory. This happens because stores and flushes to A and B are done between synchronization points. The case of C is different. Using the SFENCE instruction, we can be assured that C will always go after A and B have been flushed.

Knowing this, you can now imagine how out-of-order writes could be a problem in a program crash. If assumptions are made with respect to the order of writes between synchronization points, or if you forget to add synchronization points between writes and flushes where strict order is essential (think of a "valid flag" for a variable write, where the variable needs to be written before the flag is set to valid), you may encounter data consistency issues. Consider the pseudocode in Listing 12-37.

Listing 12-37. Pseudocode showcasing an out-of-order issue

```
1  writer () {
2          pcounter = 0;
3          flush (pcounter);
4          for (i=0; i<max; i++) {
5                  pcounter++;
6                  if (rand () % 2 == 0) {
7                          pcells[i].data = data ();
8                          flush (pcells[i].data);
9                          pcells[i].valid = True;
10                 } else {
11                         pcells[i].valid = False;
12                 }
13                 flush (pcells[i].valid);
14         }
15         flush (pcounter);
16 }
17
18 reader () {
19         for (i=0; i<pcounter; i++) {
20                 if (pcells[i].valid == True) {
21                         print (pcells[i].data);
22                 }
23         }
24 }
```

For simplicity, assume that all flushes in Listing 12-37 are also synchronization points; that is, flush() uses CLFLUSH. The logic of the program is very simple. There are two persistent memory variables: pcells and pcounter. The first is an array of tuples {data, valid} where data holds the data and valid is a flag indicating if data is valid or not. The second variable is a counter indicating how many elements in the array have been written correctly to persistent memory. In this case, the valid flag is not the one indicating whether or not the array position was written correctly to persistent memory. In this case, the flag's meaning only indicates if the function data() was called, that is, whether or not data has meaningful data.

At first glance, the program appears correct. With every new iteration of the loop, the counter is incremented, and then the array position is written and flushed. However, pcounter is incremented *before* we write to the array, thus creating a discrepancy between pcounter and the actual number of committed entries in the array. Although it is true that pcounter is not flushed until after the loop, the program is only correct after a crash if we assume that the changes to pcounter stay in the CPU caches (in that case, a program crash in the middle of the loop would simply leave the counter to zero).

As mentioned at the beginning of this section, we cannot make that assumption. A cache line can be evicted at any time. In the pseudocode example in Listing 12-37, we could run into a bug where pcounter indicates that the array is longer than it really is, making the reader() read uninitialized memory.

The code in Listings 12-38 and 12-39 provide a C++ implementation of the pseudocode from Listing 12-37. Both use libpmemobj-cpp from the PMDK. Listing 12-38 is the writer program, and Listing 12-39 is the reader.

Listing 12-38. Example of writing to persistent memory with an out-of-order write bug

```
33   #include <emmintrin.h>
34   #include <unistd.h>
35   #include <stdio.h>
36   #include <string.h>
37   #include <stdint.h>
38   #include <libpmemobj++/persistent_ptr.hpp>
39   #include <libpmemobj++/make_persistent.hpp>
40   #include <libpmemobj++/make_persistent_array.hpp>
41   #include <libpmemobj++/transaction.hpp>
42   #include <valgrind/pmemcheck.h>
43
44   using namespace std;
45   namespace pobj = pmem::obj;
46
47   struct header_t {
48       uint32_t counter;
49       uint8_t reserved[60];
50   };
```

```
51  struct record_t {
52      char name[63];
53      char valid;
54  };
55  struct root {
56      pobj::persistent_ptr<header_t> header;
57      pobj::persistent_ptr<record_t[]> records;
58  };
59
60  pobj::pool<root> pop;
61
62  int main(int argc, char *argv[]) {
63
64      // everything between BEGIN and END can be
65      // assigned a particular engine in pmreorder
66      VALGRIND_PMC_EMIT_LOG("PMREORDER_TAG.BEGIN");
67
68      pop = pobj::pool<root>::open("/mnt/pmem/file",
69                                  "RECORDS");
70      auto proot = pop.root();
71
72      // allocation of memory and initialization to zero
73      pobj::transaction::run(pop, [&] {
74          proot->header
75              = pobj::make_persistent<header_t>();
76          proot->header->counter = 0;
77          proot->records
78              = pobj::make_persistent<record_t[]>(10);
79          proot->records[0].valid = 0;
80      });
81
82      pobj::persistent_ptr<header_t> header
83          = proot->header;
84      pobj::persistent_ptr<record_t[]> records
85          = proot->records;
86
```

```
87          VALGRIND_PMC_EMIT_LOG("PMREORDER_TAG.END");
88
89          header->counter = 0;
90          for (uint8_t i = 0; i < 10; i++) {
91              header->counter++;
92              if (rand() % 2 == 0) {
93                  snprintf(records[i].name, 63,
94                           "record #%u", i + 1);
95                  pop.persist(records[i].name, 63); // flush
96                  records[i].valid = 2;
97              } else
98                  records[i].valid = 1;
99              pop.persist(&(records[i].valid), 1); // flush
100         }
101         pop.persist(&(header->counter), 4); // flush
102
103         pop.close();
104         return 0;
105     }
```

Listing 12-39. Reading the data structure written by Listing 12-38 to persistent memory

```
33  #include <stdio.h>
34  #include <stdint.h>
35  #include <libpmemobj++/persistent_ptr.hpp>
36
37  using namespace std;
38  namespace pobj = pmem::obj;
39
40  struct header_t {
41      uint32_t counter;
42      uint8_t reserved[60];
43  };
```

```
44  struct record_t {
45      char name[63];
46      char valid;
47  };
48  struct root {
49      pobj::persistent_ptr<header_t> header;
50      pobj::persistent_ptr<record_t[]> records;
51  };
52
53  pobj::pool<root> pop;
54
55  int main(int argc, char *argv[]) {
56
57      pop = pobj::pool<root>::open("/mnt/pmem/file",
58                                   "RECORDS");
59      auto proot = pop.root();
60      pobj::persistent_ptr<header_t> header
61          = proot->header;
62      pobj::persistent_ptr<record_t[]> records
63          = proot->records;
64
65      for (uint8_t i = 0; i < header->counter; i++) {
66          if (records[i].valid == 2) {
67              printf("found valid record\n");
68              printf("  name    = %s\n",
69                          records[i].name);
70          }
71      }
72
73      pop.close();
74      return 0;
75  }
```

Listing 12-38 (writer) uses the VALGRIND_PMC_EMIT_LOG macro to emit a pmreorder message when we get to lines 66 and 87. This will make sense later when we introduce out-of-order analysis using pmemcheck.

Now we will run Persistence Inspector first. To perform out-of-order analysis, we must use the -check-out-of-order-store option to the report phase. Listing 12-40 shows collecting the before and after data and then running the report.

Listing 12-40. Running Intel Inspector – Persistence Inspector with Listing 12-38 (writer) and Listing 12-39 (reader)

```
$ pmempool create obj --size=100M --layout=RECORDS /mnt/pmem/file

$ pmeminsp cb -pmem-file /mnt/pmem/file -- ./listing_12-38
++ Analysis starts

++ Analysis completes
++ Data is stored in folder "/data/.pmeminspdata/data/listing_12-38"

$ pmeminsp ca -pmem-file /mnt/pmem/file -- ./listing_12-39
++ Analysis starts

found valid record
  name    = record #2
found valid record
  name    = record #7
found valid record
  name    = record #8

++ Analysis completes
++ Data is stored in folder "/data/.pmeminspdata/data/listing_12-39"

$ pmeminsp rp -check-out-of-order-store -- ./listing_12-38 ./listing_12-39
#==============================================================
# Diagnostic # 1: Out-of-order stores
#------------------
  Memory store
    of size 4 at address 0x7FD7BEBC05D0 (offset 0x3C05D0 in /mnt/pmem/file)
    in /data/listing_12-38!main at listing_12-38.cpp:91 - 0x1D0C
    in /lib64/libc.so.6!__libc_start_main at <unknown_file>:<unknown_line>
    - 0x223D3
    in /data/listing_12-38!_start at <unknown_file>:<unknown_line> - 0x1624
```

```
is out of order with respect to

memory store
  of size 1 at address 0x7FD7BEBC068F (offset 0x3C068F in /mnt/pmem/file)
  in /data/listing_12-38!main at listing_12-38.cpp:98 - 0x1DAF
  in /lib64/libc.so.6!__libc_start_main at <unknown_file>:<unknown_line>
  - 0x223D3
  in /data/listing_12-38!_start at <unknown_file>:<unknown_line> - 0x1624
```

The Persistence Inspector report identifies an out-of-order store issue. The tool
says that incrementing the counter in line 91 (main at listing_12-38.cpp:91) is
out of order with respect to writing the valid flag inside a record in line 98 (main at
listing_12-38.cpp:98).

To perform out-of-order analysis with pmemcheck, we must introduce a new tool
called pmreorder. The pmreorder tool is included in PMDK from version 1.5 onward.
This stand-alone Python tool performs a consistency check of persistent programs
using a store reordering mechanism. The pmemcheck tool cannot do this type of analysis,
although it is still used to generate a detailed log of all the stores and flushes issued by an
application that pmreorder can parse. For example, consider Listing 12-41.

Listing 12-41. Running pmemcheck to generate a detailed log of all the stores
and flushes issued by Listing 12-38

```
$ valgrind --tool=pmemcheck -q --log-stores=yes --log-stores-
stacktraces=yes
  --log-stores-stacktraces-depth=2 --print-summary=yes
  --log-file=store_log.log ./listing_12-38
```

The meaning of each parameter is as follows:

- -q silences unnecessary pmemcheck logs that pmreorder cannot parse.

- --log-stores=yes tells pmemcheck to log all stores.

- --log-stores-stacktraces=yes dumps stacktrace with each logged
  store. This helps locate issues in your source code.

- --log-stores-stacktraces-depth=2 is the depth of logged
  stacktraces. Adjust according to the level of information you need.

- `--print-summary=yes` prints a summary on program exit. Why not?

- `--log-file=store_log.log` logs everything to `store_log.log`.

The `pmreorder` tool works with the concept of "engines." For example, the `ReorderFull` engine checks consistency for all the possible combinations of reorders of stores and flushes. This engine can be extremely slow for some programs, so you can use other engines such as `ReorderPartial` or `NoReorderDoCheck`. For more information, refer to the `pmreorder` page, which has links to the man pages (`https://pmem.io/pmdk/pmreorder/`).

Before we run `pmreorder`, we need a program that can walk the list of records contained within the memory pool and return 0 when the data structure is consistent, or 1 otherwise. This program is similar to the reader shown in Listing 12-42.

Listing 12-42. Checking the consistency of the data structure written in Listing 12-38

```
33  #include <stdio.h>
34  #include <stdint.h>
35  #include <libpmemobj++/persistent_ptr.hpp>
36
37  using namespace std;
38  namespace pobj = pmem::obj;
39
40  struct header_t {
41      uint32_t counter;
42      uint8_t reserved[60];
43  };
44  struct record_t {
45      char name[63];
46      char valid;
47  };
48  struct root {
49      pobj::persistent_ptr<header_t> header;
50      pobj::persistent_ptr<record_t[]> records;
51  };
52
```

```
53  pobj::pool<root> pop;
54
55  int main(int argc, char *argv[]) {
56
57      pop = pobj::pool<root>::open("/mnt/pmem/file",
58                                  "RECORDS");
59      auto proot = pop.root();
60      pobj::persistent_ptr<header_t> header
61          = proot->header;
62      pobj::persistent_ptr<record_t[]> records
63          = proot->records;
64
65      for (uint8_t i = 0; i < header->counter; i++) {
66          if (records[i].valid < 1 or
67                          records[i].valid > 2)
68              return 1; // data struc. corrupted
69      }
70
71      pop.close();
72      return 0; // everything ok
73  }
```

The program in Listing 12-42 iterates over all the records that we expect should have been written correctly to persistent memory (lines 65-69). It checks the valid flag for each record, which should be either 1 or 2 for the record to be correct (line 66). If an issue is detected, the checker will return 1 indicating data corruption.

Listing 12-43 shows a three-step process for analyzing the program:

1. Create an object type persistent memory pool, known as a memory-mapped file, on /mnt/pmem/file of size 100MiB, and name the internal layout "RECORDS."

2. Use the pmemcheck Valgrind tool to record data and call stacks while the program is running.

3. The pmreorder utility processes the store.log output file from pmemcheck using the ReorderFull engine to produce a final report.

Listing 12-43. First, a pool is created for Listing 12-38. Then, pmemcheck is run to get a detailed log of all the stores and flushes issued by Listing 12-38. Finally, pmreorder is run with engine ReorderFull

```
$ pmempool create obj --size=100M --layout=RECORDS /mnt/pmem/file
```

```
$ valgrind --tool=pmemcheck -q --log-stores=yes --log-stores-
stacktraces=yes --log-stores-stacktraces-depth=2 --print-summary=yes
--log-file=store.log ./listing_12-38
```

```
$ pmreorder -l store.log -o output_file.log -x PMREORDER_
TAG=NoReorderNoCheck -r ReorderFull -c prog -p ./listing_12-38
```

The meaning of each `pmreorder` option is as follows:

- `-l store_log.log` is the input file generated by `pmemcheck` with all the stores and flushes issued by the application.

- `-o output_file.log` is the output file with the out-of-order analysis results.

- `-x PMREORDER_TAG=NoReorderNoCheck` assigns the engine `NoReorderNoCheck` to the code enclosed by the tag `PMREORDER_TAG` (see lines 66-87 from Listing 12-38). This is done to focus the analysis on the loop only (lines 89-105 from Listing 12-38).

- `-r ReorderFull` sets the initial reorder engine. In our case, `ReorderFull`.

- `-c prog` is the consistency checker type. It can be `prog` (program) or `lib` (library).

- `-p ./checker` is the consistency checker.

Opening the generated file `output_file.log`, you should see entries similar to those in Listing 12-44 that highlight detected inconsistencies and problems within the code.

Listing 12-44. Content from "output_file.log" generated by pmreorder showing a detected inconsistency during the out-of-order analysis

```
WARNING:pmreorder:File /mnt/pmem/file inconsistent
WARNING:pmreorder:Call trace:
Store [0]:
    by  0x401D0C: main (listing_12-38.cpp:91)
```

The report states that the problem resides at line 91 of the listing_12-38.cpp writer program. To fix listing_12-38.cpp, move the counter incrementation after all the data in the record has been flushed all the way to persistent media. Listing 12-45 shows the corrected part of the code.

Listing 12-45. Fix Listing 12-38 by moving the incrementation of the counter to the end of the loop (line 95)

```
86     for (uint8_t i = 0; i < 10; i++) {
87         if (rand() % 2 == 0) {
88             snprintf(records[i].name, 63,
89                     "record #%u", i + 1);
90             pop.persist(records[i].name, 63);
91             records[i].valid = 2;
92         } else
93             records[i].valid = 1;
94         pop.persist(&(records[i].valid), 1);
95         header->counter++;
96     }
```

Summary

This chapter provided an introduction to each tool and described how to use them. Catching issues early in the development cycle can save countless hours of debugging complex code later on. This chapter introduced three valuable tools – Persistence Inspector, pmemcheck, and pmreorder – that persistent memory programmers will want to integrate into their development and testing cycles to detect issues. We demonstrated how useful these tools are at detecting many different types of common programming errors.

The Persistent Memory Development Kit (PMDK) uses the tools described here to ensure each release is fully validated before it is shipped. The tools are tightly integrated into the PMDK continuous integration (CI) development cycle, so you can quickly catch and fix issues.

CHAPTER 13

Enabling Persistence Using a Real-World Application

This chapter turns the theory from Chapter 4 (and other chapters) into practice. We show how an application can take advantage of persistent memory by building a persistent memory-aware database storage engine. We use MariaDB (`https://mariadb.org/`), a popular open source database, as it provides a pluggable storage engine model. The completed storage engine is not intended for production use and does not implement all the features a production quality storage engine should. We implement only the basic functionality to demonstrate how to begin persistent memory programming using a well known database. The intent is to provide you with a more hands-on approach for persistent memory programming so you may enable persistent memory features and functionality within your own application. Our storage engine is left as an optional exercise for you to complete. Doing so would create a new persistent memory storage engine for MariaDB, MySQL, Percona Server, and other derivatives. You may also choose to modify an existing MySQL database storage engine to add persistent memory features, or perhaps choose a different database entirely.

We assume that you are familiar with the preceding chapters that covered the fundamentals of the persistent memory programming model and Persistent Memory Development Kit (PMDK). In this chapter, we implement our storage engine using C++ and `libpmemobj-cpp` from Chapter 8. If you are not a C++ developer, you will still find this information helpful because the fundamentals apply to other languages and applications.

The complete source code for the persistent memory-aware database storage engine can be found on GitHub at `https://github.com/pmem/pmdk-examples/tree/master/pmem-mariadb`.

© The Author(s) 2020
S. Scargall, *Programming Persistent Memory*, https://doi.org/10.1007/978-1-4842-4932-1_13

The Database Example

A tremendous number of existing applications can be categorized in many ways. For the purpose of this chapter, we explore applications from the common components perspective, including an interface, a business layer, and a store. The interface interacts with the user, the business layer is a tier where the application's logic is implemented, and the store is where data is kept and processed by the application.

With so many applications available today, choosing one to include in this book that would satisfy all or most of our requirements was difficult. We chose to use a database as an example because a unified way of accessing data is a common denominator for many applications.

Different Persistent Memory Enablement Approaches

The main advantages of persistent memory include:

- It provides access latencies that are lower than flash SSDs.

- It has higher throughput than NAND storage devices.

- Real-time access to data allows ultrafast access to large datasets.

- Data persists in memory after a power interruption.

Persistent memory can be used in a variety of ways to deliver lower latency for many applications:

- **In-memory databases:** In-memory databases can leverage persistent memory's larger capacities and significantly reduce restart times. Once the database memory maps the index, tables, and other files, the data is immediately accessible. This avoids lengthy startup times where the data is traditionally read from disk and paged in to memory before it can be accessed or processed.

- **Fraud detection:** Financial institutions and insurance companies can perform real-time data analytics on millions of records to detect fraudulent transactions.

- **Cyber threat analysis:** Companies can quickly detect and defend against increasing cyber threats.

- **Web-scale personalization:** Companies can tailor online user experiences by returning relevant content and advertisements, resulting in higher user click-through rate and more e-commerce revenue opportunities.

- **Financial trading:** Financial trading applications can rapidly process and execute financial transactions, allowing them to gain a competitive advantage and create a higher revenue opportunity.

- **Internet of Things (IoT):** Faster data ingest and processing of huge datasets in real-time reduces time to value.

- **Content delivery networks (CDN):** A CDN is a highly distributed network of edge servers strategically placed across the globe with the purpose of rapidly delivering digital content to users. With a memory capacity, each CDN node can cache more data and reduce the total number of servers, while networks can reliably deliver low-latency data to their clients. If the CDN cache is persisted, a node can restart with a warm cache and sync only the data it is missed while it was out of the cluster.

Developing a Persistent Memory-Aware MariaDB* Storage Engine

The storage engine developed here is not production quality and does not implement all the functionality expected by most database administrators. To demonstrate the concepts described earlier, we kept the example simple, implementing table `create()`, `open()`, and `close()` operations and INSERT, UPDATE, DELETE, and SELECT SQL operations. Because the storage engine capabilities are quite limited without indexing, we include a simple indexing system using volatile memory to provide faster access to the data residing in persistent memory.

Although MariaDB has many storage engines to which we could add persistent memory, we are building a new storage engine from scratch in this chapter. To learn more about the MariaDB storage engine API and how storage engines work, we suggest reading the MariaDB "Storage Engine Development" documentation (`https://mariadb.com/kb/en/library/storage-engines-storage-engine-development/`). Since MariaDB is based on MySQL, you can also refer to the MySQL "Writing a Custom

Storage Engine" documentation (`https://dev.mysql.com/doc/internals/en/custom-engine.html`) to find all the information for creating an engine from scratch.

Understanding the Storage Layer

MariaDB provides a pluggable architecture for storage engines that makes it easier to develop and deploy new storage engines. A pluggable storage engine architecture also makes it possible to create new storage engines and add them to a running MariaDB server without recompiling the server itself. The storage engine provides data storage and index management for MariaDB. The MariaDB server communicates with the storage engines through a well-defined API.

In our code, we implement a prototype of a pluggable persistent memory–enabled storage engine for MariaDB using the `libpmemobj` library from the Persistent Memory Development Kit (PMDK).

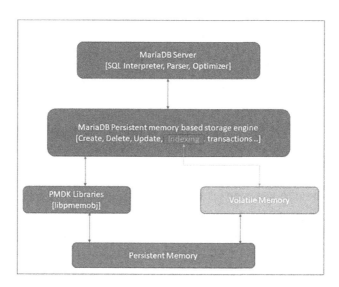

Figure 13-1. *MariaDB storage engine architecture diagram for persistent memory*

Figure 13-1 shows how the storage engine communicates with `libpmemobj` to manage the data stored in persistent memory. The library is used to turn a persistent memory pool into a flexible object store.

Creating a Storage Engine Class

The implementation of the storage engine described here is single-threaded to support a single session, a single user, and single table requests. A multi-threaded implementation would detract from the focus of this chapter. Chapter 14 discussed concurrency in more detail. The MariaDB server communicates with storage engines through a well-defined handler interface that includes a `handlerton`, which is a singleton `handler` that is connected to a table handler. The `handlerton` defines the storage engine and contains pointers to the methods that apply to the persistent memory storage engine.

The first method the storage engine needs to support is to enable the call for a new handler instance, shown in Listing 13-1.

Listing 13-1. ha_pmdk.cc – Creating a new handler instance

```
117   static handler *pmdk_create_handler(handlerton *hton,
118                                       TABLE_SHARE *table,
119                                       MEM_ROOT *mem_root);
120
121   handlerton *pmdk_hton;
```

When a `handler` instance is created, the MariaDB server sends commands to the `handler` to perform data storage and retrieve tasks such as opening a table, manipulating rows, managing indexes, and transactions. When a `handler` is instantiated, the first required operation is the opening of a table. Since the storage engine is a single user and single-threaded implementation, only one `handler` instance is created.

Various `handler` methods are also implemented; they apply to the storage engine as a whole, as opposed to methods like `create()` and `open()` that work on a per-table basis. Some examples of such methods include transaction methods to handle commits and rollbacks, shown in Listing 13-2.

Listing 13-2. ha_pmdk.cc – Handler methods including transactions, rollback, etc

```
209   static int pmdk_init_func(void *p)
210   {
...
213     pmdk_hton= (handlerton *)p;
214     pmdk_hton->state=    SHOW_OPTION_YES;
215     pmdk_hton->create=  pmdk_create_handler;
```

```
216     pmdk_hton->flags=    HTON_CAN_RECREATE;
217     pmdk_hton->tablefile_extensions= ha_pmdk_exts;
218
219     pmdk_hton->commit= pmdk_commit;
220     pmdk_hton->rollback= pmdk_rollback;
...
223  }
```

The abstract methods defined in the handler class are implemented to work with persistent memory. An internal representation of the objects in persistent memory is created using a single linked list (SLL). This internal representation is very helpful to iterate through the records to improve performance.

To perform a variety of operations and gain faster and easier access to data, we used the simple row structure shown in Listing 13-3 to hold the pointer to persistent memory and the associated field value in the buffer.

Listing 13-3. ha_pmdk.h – A simple data structure to store data in a single linked list

```
71   struct row {
72     persistent_ptr<row> next;
73     uchar buf[];
74   };
```

Creating a Database Table

The create() method is used to create the table. This method creates all necessary files in persistent memory using libpmemobj. As shown in Listing 13-4, we create a new pmemobj type pool for each table using the pmemobj_create() method; this method creates a transactional object store with the given total poolsize. The table is created in the form of an .obj extension.

Listing 13-4. Creating a table method

```
1247  int ha_pmdk::create(const char *name, TABLE *table_arg,
1248                       HA_CREATE_INFO *create_info)
1249  {
1250
```

```
1251    char path[MAX_PATH_LEN];
1252    DBUG_ENTER("ha_pmdk::create");
1253    DBUG_PRINT("info", ("create"));
1254
1255    snprintf(path, MAX_PATH_LEN, "%s%s", name, PMEMOBJ_EXT);
1256    PMEMobjpool *pop = pmemobj_create(path, name,PMEMOBJ_MIN_POOL,
        S_IRWXU);
1257    if (pop == NULL) {
1258      DBUG_PRINT("info", ("failed : %s error number :
          %d",path,errCodeMap[errno]));
1259      DBUG_RETURN(errCodeMap[errno]);
1260    }
1261    DBUG_PRINT("info", ("Success"));
1262    pmemobj_close(pop);
1263
1264    DBUG_RETURN(0);
1265  }
```

Opening a Database Table

Before any read or write operations are performed on a table, the MariaDB server calls
the open()method to open the data and index tables. This method opens all the named
tables associated with the persistent memory storage engine at the time the storage
engine starts. A new table class variable, objtab, was added to hold the PMEMobjpool.
The names for the tables to be opened are provided by the MariaDB server. The index
container in volatile memory is populated using the open() function call at the time of
server start using the loadIndexTableFromPersistentMemory() function.

The pmemobj_open() function from libpmemobj is used to open an existing object
store memory pool (see Listing 13-5). The table is also opened at the time of a table
creation if any read/write action is triggered.

Listing 13-5. ha_pmdk.cc – Opening a database table

```
290  int ha_pmdk::open(const char *name, int mode, uint test_if_locked)
291  {
...
```

```
302    objtab = pmemobj_open(path, name);
303    if (objtab == NULL)
304      DBUG_RETURN(errCodeMap[errno]);
305
306    proot = pmemobj_root(objtab, sizeof (root));
307    // update the MAP when start occured
308    loadIndexTableFromPersistentMemory();
...
310  }
```

Once the storage engine is up and running, we can begin to insert data into it. But we first must implement the INSERT, UPDATE, DELETE, and SELECT operations.

Closing a Database Table

When the server is finished working with a table, it calls the closeTable() method to close the file using pmemobj_close() and release any other resources (see Listing 13-6). The pmemobj_close() function closes the memory pool indicated by objtab and deletes the memory pool handle.

Listing 13-6. ha_pmdk.cc – Closing a database table

```
376  int ha_pmdk::close(void)
377  {
378    DBUG_ENTER("ha_pmdk::close");
379    DBUG_PRINT("info", ("close"));
380
381    pmemobj_close(objtab);
382    objtab = NULL;
383
384    DBUG_RETURN(0);
385  }
```

INSERT Operation

The INSERT operation is implemented in the write_row() method, shown in Listing 13-7. During an INSERT, the row objects are maintained in a singly linked list. If the table is indexed, the index table container in volatile memory is updated with the new row objects after the persistent operation completes successfully. write_row() is an important method because, in addition to the allocation of persistent pool storage to the rows, it is used to populate the indexing containers. pmemobj_tx_alloc() is used for inserts. write_row() transactionally allocates a new object of a given size and type_num.

Listing 13-7. ha_pmdk.cc – Closing a database table

```
417   int ha_pmdk::write_row(uchar *buf)
418   {
...
421     int err = 0;
422
423     if (isPrimaryKey() == true)
424       DBUG_RETURN(HA_ERR_FOUND_DUPP_KEY);
425
426     persistent_ptr<row> row;
427     TX_BEGIN(objtab) {
428       row = pmemobj_tx_alloc(sizeof (row) + table->s->reclength, 0);
429       memcpy(row->buf, buf, table->s->reclength);
430       row->next = proot->rows;
431       proot->rows = row;
432     } TX_ONABORT {
433       DBUG_PRINT("info", ("write_row_abort errno :%d ",errno));
434       err = errno;
435     } TX_END
436     stats.records++;
437
438     for (Field **field = table->field; *field; field++) {
439       if ((*field)->key_start.to_ulonglong() >= 1) {
440         std::string convertedKey = IdentifyTypeAndConvertToString((*fie
ld)->ptr, (*field)->type(),(*field)->key_length(),1);
441         insertRowIntoIndexTable(*field, convertedKey, row);
442       }
443     }
444     DBUG_RETURN(err);
445   }
```

In every INSERT operation, the field values are checked for a preexisting duplicate. The primary key field in the table is checked using the isPrimaryKey()function (line 423). If the key is a duplicate, the error HA_ERR_FOUND_DUPP_KEY is returned. The isPrimaryKey() is implemented in Listing 13-8.

Listing 13-8. ha_pmdk.cc – Checking for duplicate primary keys

```
462  bool ha_pmdk::isPrimaryKey(void)
463  {
464    bool ret = false;
465    database *db = database::getInstance();
466    table_ *tab;
467    key *k;
468    for (unsigned int i= 0; i < table->s->keys; i++) {
469      KEY* key_info = &table->key_info[i];
470      if (memcmp("PRIMARY",key_info->name.str,sizeof("PRIMARY"))==0) {
471        Field *field = key_info->key_part->field;
472        std::string convertedKey = IdentifyTypeAndConvertToString
                 (field->ptr, field->type(),field->key_length(),1);
473        if (db->getTable(table->s->table_name.str, &tab)) {
474          if (tab->getKeys(field->field_name.str, &k)) {
475            if (k->verifyKey(convertedKey)) {
476              ret = true;
477              break;
478            }
479          }
480        }
481      }
482    }
483    return ret;
484  }
```

UPDATE Operation

The server executes UPDATE statements by performing a rnd_init() or index_init() table scan until it locates a row matching the key value in the WHERE clause of the UPDATE statement before calling the update_row() method. If the table is an indexed table, the index container is also updated after this operation is successful. In the update_row() method defined in Listing 13-9, the old_data field will have the previous row record in it, while new_data will have the new data.

Listing 13-9. ha_pmdk.cc – Updating existing row data

```
506  int ha_pmdk::update_row(const uchar *old_data, const uchar *new_data)
507  {
...
540              if (k->verifyKey(key_str))
541                k->updateRow(key_str, field_str);
...
551    if (current)
552      memcpy(current->buf, new_data, table->s->reclength);
...
```

The index table is also updated using the updateRow() method shown in Listing 13-10.

Listing 13-10. ha_pmdk.cc – Updating existing row data

```
1363  bool key::updateRow(const std::string oldStr, const std::string newStr)
1364  {
...
1366    persistent_ptr<row> row_;
1367    bool ret = false;
1368    rowItr matchingEleIt = getCurrent();
1369
1370    if (matchingEleIt->first == oldStr) {
1371      row_ = matchingEleIt->second;
1372      std::pair<const std::string, persistent_ptr<row> > r(newStr, row_);
1373      rows.erase(matchingEleIt);
1374      rows.insert(r);
1375      ret = true;
1376    }
1377    DBUG_RETURN(ret);
1378  }
```

DELETE Operation

The DELETE operation is implemented using the delete_row() method. Three different
scenarios should be considered:

- Deleting an indexed value from the indexed table

- Deleting a non-indexed value from the indexed table

- Deleting a field from the non-indexed table

For each scenario, different functions are called. When the operation is successful,
the entry is removed from both the index (if the table is an indexed table) and persistent
memory. Listing 13-11 shows the logic to implement the three scenarios.

Listing 13-11. ha_pmdk.cc – Updating existing row data

```
594  int ha_pmdk::delete_row(const uchar *buf)
595  {
...
602    // Delete the field from non indexed table
603    if (active_index == 64 && table->s->keys ==0 ) {
604      if (current)
605        deleteNodeFromSLL();
606    } else if (active_index == 64 && table->s->keys !=0 ) { // Delete
       non indexed column field from indexed table
607      if (current) {
608        deleteRowFromAllIndexedColumns(current);
609        deleteNodeFromSLL();
610      }
611    } else { // Delete indexed column field from indexed table
612    database *db = database::getInstance();
613    table_ *tab;
614    key *k;
615    KEY_PART_INFO *key_part = table->key_info[active_index].key_part;
616    if (db->getTable(table->s->table_name.str, &tab)) {
617        if (tab->getKeys(key_part->field->field_name.str, &k)) {
618          rowItr currNode = k->getCurrent();
619          rowItr prevNode = std::prev(currNode);
```

```
620            if (searchNode(prevNode->second)) {
621              if (prevNode->second) {
622                deleteRowFromAllIndexedColumns(prevNode->second);
623                deleteNodeFromSLL();
624              }
625            }
626          }
627        }
628      }
629    stats.records--;
630
631    DBUG_RETURN(0);
632  }
```

Listing 13-12 shows how the deleteRowFromAllIndexedColumns() function deletes the value from the index containers using the deleteRow() method.

Listing 13-12. ha_pmdk.cc – Deletes an entry from the index containers

```
634  void ha_pmdk::deleteRowFromAllIndexedColumns(const persistent_ptr<row>
     &row)
635  {
...
643      if (db->getTable(table->s->table_name.str, &tab)) {
644        if (tab->getKeys(field->field_name.str, &k)) {
645          k->deleteRow(row);
646        }
...
```

The deleteNodeFromSLL() method deletes the object from the linked list residing on persistent memory using libpmemobj transactions, as shown in Listing 13-13.

Listing 13-13. ha_pmdk.cc – Deletes an entry from the linked list using transactions

```
651  int ha_pmdk::deleteNodeFromSLL()
652  {
653    if (!prev) {
654      if (!current->next) { // When sll contains single node
655        TX_BEGIN(objtab) {
656          delete_persistent<row>(current);
657          proot->rows = nullptr;
658        } TX_END
659      } else { // When deleting the first node of sll
660        TX_BEGIN(objtab) {
661          delete_persistent<row>(current);
662          proot->rows = current->next;
663          current = nullptr;
664        } TX_END
665      }
666    } else {
667      if (!current->next) { // When deleting the last node of sll
668        prev->next = nullptr;
669      } else { // When deleting other nodes of sll
670        prev->next = current->next;
671      }
672      TX_BEGIN(objtab) {
673        delete_persistent<row>(current);
674        current = nullptr;
675      } TX_END
676    }
677    return 0;
678  }
```

SELECT Operation

SELECT is an important operation that is required by several methods. Many methods that are implemented for the SELECT operation are also called from other methods. The rnd_init() method is used to prepare for a table scan for non-indexed tables, resetting counters and pointers to the start of the table. If the table is an indexed table, the MariaDB server calls the index_init() method. As shown in Listing 13-14, the pointers are initialized.

Listing 13-14. ha_pmdk.cc – rnd_init() is called when the system wants the storage engine to do a table scan

```
869  int ha_pmdk::rnd_init(bool scan)
870  {
...
874    current=prev=NULL;
875    iter = proot->rows;
876    DBUG_RETURN(0);
877  }
```

When the table is initialized, the MariaDB server calls the rnd_next(), index_first(), or index_read_map() method, depending on whether the table is indexed or not. These methods populate the buffer with data from the current object and updates the iterator to the next value. The methods are called once for every row to be scanned.

Listing 13-15 shows how the buffer passed to the function is populated with the contents of the table row in the internal MariaDB format. If there are no more objects to read, the return value must be HA_ERR_END_OF_FILE.

Listing 13-15. ha_pmdk.cc – rnd_init() is called when the system wants the storage engine to do a table scan

```
902  int ha_pmdk::rnd_next(uchar *buf)
903  {
...
910    memcpy(buf, iter->buf, table->s->reclength);
911    if (current != NULL) {
912      prev = current;
913    }
```

```
914     current = iter;
915     iter = iter->next;
916
917     DBUG_RETURN(0);
918 }
```

This concludes the basic functionality our persistent memory enabled storage engine set out to achieve. We encourage you to continue the development of this storage engine to introduce more features and functionality.

Summary

This chapter provided a walk-through using `libpmemobj` from the PMDK to create a persistent memory-aware storage engine for the popular open source MariaDB database. Using persistent memory in an application can provide continuity in the event of an unplanned system shutdown along with improved performance gained by storing your data close to the CPU where you can access it at the speed of the memory bus. While database engines commonly use in-memory caches for performance, which take time to warm up, persistent memory offers an immediately warm cache upon application startup.

CHAPTER 14

Concurrency and Persistent Memory

This chapter discusses what you need to know when building multithreaded applications for persistent memory. We assume you already have experience with multithreaded programming and are familiar with basic concepts such as mutexes, critical section, deadlocks, atomic operations, and so on.

The first section of this chapter highlights common practical solutions for building multithreaded applications for persistent memory. We describe the limitation of the Persistent Memory Development Kit (PMDK) transactional libraries, such as libpmemobj and libpmemobj-cpp, for concurrent execution. We demonstrate simple examples that are correct for volatile memory but cause data inconsistency issues on persistent memory in situations where the transaction aborts or the process crashes. We also discuss why regular mutexes cannot be placed as is on persistent memory and introduce the persistent deadlock term. Finally, we describe the challenges of building lock-free algorithms for persistent memory and continue our discussion of visibility vs. persistency from previous chapters.

The second section demonstrates our approach to designing concurrent data structures for persistent memory. At the time of publication, we have two concurrent associative C++ data structures developed for persistent memory - a concurrent hash map and a concurrent map. More will be added over time. We discuss both implementations within this chapter.

All code samples are implemented in C++ using the libpmemobj-cpp library described in Chapter 8. In this chapter, we usually refer to libpmemobj because it implements the features and libpmemobj-cpp is only a C++ extension wrapper for it. The concepts are general and can apply to any programming language.

© The Author(s) 2020
S. Scargall, *Programming Persistent Memory*, https://doi.org/10.1007/978-1-4842-4932-1_14

Transactions and Multithreading

In computer science, ACID (atomicity, consistency, isolation, and durability) is a set of properties of transactions intended to guarantee data validity and consistency in case of errors, power failures, and abnormal termination of a process. Chapter 7 introduced PMDK transactions and their ACID properties. This chapter focuses on the relevancy of multithreaded programs for persistent memory. Looking forward, Chapter 16 will provide some insights into the internals of libpmemobj transactions.

The small program in Listing 14-1 shows that the counter stored within the root object is incremented concurrently by multiple threads. The program opens the persistent memory pool and prints the value of counter. It then runs ten threads, each of which calls the increment() function. Once all the threads complete successfully, the program prints the final value of counter.

Listing 14-1. Example to demonstrate that PMDK transactions do not automatically support isolation

```
41  using namespace std;
42  namespace pobj = pmem::obj;
43
44  struct root {
45      pobj::p<int> counter;
46  };
47
48  using pop_type = pobj::pool<root>;
49
50  void increment(pop_type &pop) {
51      auto proot = pop.root();
52      pobj::transaction::run(pop, [&] {
53          proot->counter.get_rw() += 1;
54      });
55  }
56
57  int main(int argc, char *argv[]) {
58      pop_type pop =
59          pop_type::open("/pmemfs/file", "COUNTER_INC");
60
```

```
61      auto proot = pop.root();
62
63      cout << "Counter = " << proot->counter << endl;
64
65      std::vector<std::thread> workers;
66      workers.reserve(10);
67      for (int i = 0; i < 10; ++i) {
68          workers.emplace_back(increment, std::ref(pop));
69      }
70
71      for (int i = 0; i < 10; ++i) {
72          workers[i].join();
73      }
74
75      cout << "Counter = " << proot->counter << endl;
76
77      pop.close();
78      return 0;
79  }
```

You might expect that the program in Listing 14-1 the prints a final counter value
of 10. However, PMDK transactions do not automatically support isolation from the
ACID properties set. The result of the increment operation on line 53 is visible to
other concurrent transactions before the current transaction has implicitly committed
its update on line 54. That is, a simple data race is occurring in this example. A race
condition occurs when two or more threads can access shared data and they try to
change it at the same time. Because the operating system's thread scheduling algorithm
can swap between threads at any time, there is no way for the application to know the
order in which the threads will attempt to access the shared data. Therefore, the result
of the change of the data is dependent on the thread scheduling algorithm, that is, both
threads are "racing" to access/change the data.

If we run this example multiple times, the results will vary from run to run. We can
try to fix the race condition by acquiring a mutex lock before the counter increment as
shown in Listing 14-2.

Listing 14-2. Example of incorrect synchronization inside a PMDK transaction

```
46   struct root {
47       pobj::mutex mtx;
48       pobj::p<int> counter;
49   };
50
51   using pop_type = pobj::pool<root>;
52
53   void increment(pop_type &pop) {
54       auto proot = pop.root();
55       pobj::transaction::run(pop, [&] {
56           std::unique_lock<pobj::mutex> lock(proot->mtx);
57           proot->counter.get_rw() += 1;
58       });
59   }
```

- Line 47: We added a `mutex` to the root data structure.

- Line 56: We acquired the mutex lock within the transaction before incrementing the value of `counter` to avoid a race condition. Each thread increments the counter inside the critical section protected by the mutex.

Now if we run this example multiple times, it will always increment the value of the `counter` stored in persistent memory by 1. But we are not done yet. Unfortunately, the example in Listing 14-2 is also wrong and can cause data inconsistency issues on persistent memory. The example works well if there are no transaction aborts. However, if the transaction aborts after the lock is released but before the transaction has completed and successfully committed its update to persistent memory, other threads can read a cached value of the counter that can cause data inconsistency issues. To understand the problem, you need to know how `libpmemobj` transactions work internally. For now, we discuss only the necessary details required to understand this issue and leave the in-depth discussion of transactions and their implementation for Chapter 16.

A `libpmemobj` transaction guarantees atomicity by tracking changes in the undo log. In the case of a failure or transaction abort, the old values for uncommitted changes are restored from the undo log. It is important to know that the undo log is a thread-specific entity. This means that each thread has its own undo log that is not synchronized with undo logs of other threads.

Figure 14-1 illustrates the internals of what happens within the transaction when we call the `increment()` function in Listing 14-2. For illustrative purposes, we only describe two threads. Each thread executes concurrent transactions to increment the value of `counter` allocated in persistent memory. We assume the initial value of `counter` is 0 and the first thread acquires the lock, while the second thread waits on the lock. Inside the critical section, the first thread adds the initial value of `counter` to the undo log and increments it. The mutex is released when execution flow leaves the lambda scope, but the transaction has not committed the update to persistent memory. The changes become immediately visible to the second thread. After a user-provided lambda is executed, the transaction needs to flush all changes to persistent memory to mark the change(s) as committed. Concurrently, the second thread adds the current value of `counter`, which is now 1, to its undo log and performs the increment operation. At that moment, there are two uncommitted transactions. The undo log of Thread 1 contains `counter = 0`, and the undo log of Thread 2 contains `counter = 1`. If Thread 2 commits its transaction while Thread 1 aborts its transaction for some reason (crash or abort), the incorrect value of `counter` will be restored from the undo log of Thread 1.

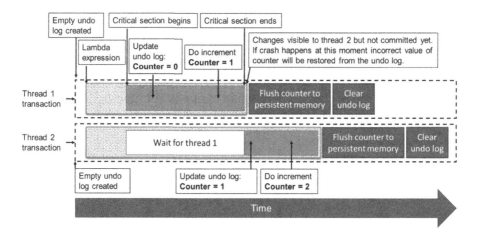

***Figure 14-1.** Illustrative execution of the Listing 14-2 example*

The solution is to hold the mutex until the transaction is fully committed, and the data has been successfully flushed to persistent memory. Otherwise, changes made by one transaction become visible to concurrent transactions before it is persisted and committed. Listing 14-3 demonstrates how to implement the `increment()` function correctly.

Listing 14-3. Correct example for concurrent PMDK transaction

```
52  void increment(pop_type &pop) {
53      auto proot = pop.root();
54      pobj::transaction::run(pop, [&] {
55          proot->counter.get_rw() += 1;
56      }, proot->mtx);
57  }
```

The `libpmemobj` API allows us to specify locks that should be acquired and held for the entire duration of the transaction. In the Listing 14-3 example, we pass the `proot->mtx` mutex object to the `run()` method as a third parameter.

Mutexes on Persistent Memory

Our previous examples used `pmem::obj::mutex` as a type for the `mtx` member in our `root` data structure instead of the regular `std::mutex` provided by Standard Template Library. The `mtx` object is a member of the `root` object that resides in persistent memory. The `std::mutex` type cannot be used on persistent memory because it may cause persistent deadlock.

A persistent deadlock happens if an application crash occurs while holding a mutex. When the program starts, if it does not release or reinitialize the mutex at startup, threads that try to acquire it will wait forever. To avoid such situations, `libpmemobj` provides synchronization primitives that reside in persistent memory. The main feature of synchronization primitives is that they are automatically reinitialized every time the persistent object store pool is open.

For C++ developers, the `libpmemobj-cpp` library provides C++11-like synchronization primitives shown in Table 14-1.

Table 14-1. *Synchronization primitives provided by libpmemob++ library*

Class	Description
pmem::obj::mutex	This class is an implementation of a persistent memory resident mutex which mimics in behavior the C++11 std::mutex. This class satisfies all requirements of the Mutex and StandardLayoutType concepts.
pmem::obj::timed_mutex	This class is an implementation of a persistent memory resident timed_mutex which mimics in behavior the C++11 std::timed_mutex. This class satisfies all requirements of TimedMutex and StandardLayoutType concepts.
pmem::obj::shared_mutex	This class is an implementation of a persistent memory resident shared_mutex which mimics in behavior the C++17 std::shared_mutex. This class satisfies all requirements of SharedMutex and StandardLayoutType concepts.
pmem::obj:: condition_variable	This class is an implementation of a persistent memory resident condition variable which mimics in behavior the C++11 std::condition_variable. This class satisfies all requirements of StandardLayoutType concept.

For C developers, the `libpmemobj` library provides pthread-like synchronization primitives shown in Table 14-2. Persistent memory-aware locking implementations are based on the standard POSIX Thread Library and provide semantics similar to standard pthread locks.

Table 14-2. *Synchronization primitives provided by the libpmemobj library*

Structure	Description
PMEMmutex	The data structure represents a persistent memory resident mutex similar to `pthread_mutex_t`.
PMEMrwlock	The data structure represents a persistent memory resident read-write lock similar to `pthread_rwlock_t`.
PMEMcond	The data structure represents a persistent memory resident condition variable similar to `pthread_cond_t`.

These convenient persistent memory-aware synchronization primitives are available for C and C++ developers. But what if a developer wants to use a custom synchronization object that is more appropriate for a particular use case? As we mentioned earlier, the main feature of persistent memory-aware synchronization primitives is that they are reinitialized every time we open a persistent memory pool. The libpmemobj-cpp library provides a more generic mechanism to reinitialize any user-provided type every time a persistent memory pool is opened.

The libpmemobj-cpp provides the pmem::obj::v<T> class template which allows creating a volatile field inside a persistent data structure. The mutex object is semantically a volatile entity, and the state of a mutex should not survive an application restart. On application restart, a mutex object should be in the unlocked state. The pmem::obj::v<T> class template is targeted for this purpose. Listing 14-4 demonstrates how to use the pmem::obj::v<T> class template with std::mutex on persistent memory.

Listing 14-4. Example demonstrating usage of std::mutex on persistent memory

```
38   namespace pobj = pmem::obj;
39
40   struct root {
41       pobj::experimental::v<std::mutex> mtx;
42   };
43
44   using pop_type = pobj::pool<root>;
45
46   int main(int argc, char *argv[]) {
47       pop_type pop =
48           pop_type::open("/pmemfs/file", "MUTEX");
49
50       auto proot = pop.root();
51
52       proot->mtx.get().lock();
53
54       pop.close();
55       return 0;
56   }
```

- Line 41: We are only storing the `mtx` object inside root object on persistent memory.

- Lines 47-48: We open the persistent memory pool with the layout name of "MUTEX".

- Line 50: We obtain a pointer to the root data structure within the pool.

- Line 52: We acquire the mutex.

- Lines 54-56: Close the pool and exit the program.

As you can see, we do not explicitly unlock the mutex within the `main()` function. If we run this example several times, the `main()` function can always lock the mutex on line 52. This works because the `pmem::obj::v<T>` class template implicitly calls a default constructor, which is a wrapped `std::mutex` object type. The constructor is called every time we open the persistent memory pool so we never run into a situation where the lock is already acquired.

If we change the `mtx` object type on line 41 from `pobj::experimental::v<std::mu tex>` to `std::mutex` and try to run the program again, the example will hang during the second run on line 52 because `mtx` object was locked during the first run and we never released it.

Atomic Operations and Persistent Memory

Atomic operations cannot be used inside PMDK transactions for the reason described in Figure 14-1. Changes made by atomic operations inside a transaction become visible to other concurrent threads before the transaction is committed. It forces data inconsistency issues in cases of abnormal program termination or transaction aborts. Consider lock-free algorithms where concurrency is achieved by atomically updating the state in memory.

Lock-Free Algorithms and Persistent Memory

It is intuitive to think that lock-free algorithms are naturally fit for persistent memory. In lock-free algorithms, thread-safety is achieved by atomic transitions between consistent states, and this is exactly what we need to support data consistency in persistent memory. But this assumption is not always correct.

To understand the problem with lock-free algorithms, remember that a system with persistent memory will usually have the virtual memory subsystem divided into two domains: volatile and persistent (described in Chapter 2). The result of an atomic operation may only update data in a CPU cache using a cache coherency protocol. There is no guarantee that the data will be flushed unless an explicit flush operation is called. CPU caches are only included within the persistence domain on platforms with eADR support. This is not mandatory for persistent memory. ADR is the minimal platform requirement for persistent memory, and in that case, CPU caches are not flushed in a power failure.

Figure 14-2 assumes a system with ADR support. The example shows concurrent lock-free insert operations to a singly linked list located in persistent memory. Two threads are trying to insert new nodes to the tail of a linked list using a compare-and-exchange (CMPXCHG instruction) operation followed by a cache flush operation (CLWB instruction). Assume Thread 1 succeeds with its compare-and-exchange, so the change appears in a volatile domain and becomes visible to the second thread. At this moment, Thread 1 may be preempted (changes not flushed to a persistent domain), while Thread 2 inserts Node 5 after Node 4 and flushes it to a persistent domain. A possibility for data inconsistency exists because Thread 2 performed an update based on the data that is not yet persisted by Thread 1.

Figure 14-2. *Example of a concurrent lock-free insert operation to a singly linked list located in persistent memory*

Concurrent Data Structures for Persistent Memory

This section describes two concurrent data structures available in the libpmemobj-cpp library: pmem::obj::concurrent_map and pmem::obj::concurrent_hash_map. Both are associative data structures composed of a collection of key and value pairs, such that each possible key appears at most once in the collection. The main difference between them is that the concurrent hash map is unordered, while the concurrent map is ordered by keys.

We define *concurrent* in this context to be the method of organizing data structures for access by multiple threads. Such data structures are intended for use in a parallel computing environment when multiple threads can concurrently call methods of a data structure without additional synchronization required.

C++ Standard Template Library (STL) data structures can be wrapped in a coarse-grained mutex to make them safe for concurrent access by letting only one thread operate on the container at a time. However, that approach eliminates concurrency and thereby restricts parallel speedup if implemented in performance-critical code. Designing concurrent data structures is a challenging task. The difficulty increases significantly when we need to develop concurrent data structures for persistent memory and make them fault tolerant.

The pmem::obj::concurrent_map and pmem::obj::concurrent_hash_map structures were inspired by the Intel Threading Building Blocks (Intel TBB),[1] which provides implementations of these concurrent data structures designed for volatile memory. You can read the *Pro TBB: C++ Parallel Programming with Threading Building* Blocks book[2] to get more information and learn how to use these concurrent data structures in your application. The free electronic copy is available from Apress at https://www.apress.com/gp/book/9781484243978.

There are three main methods in our concurrent associative data structures: find, insert, and erase/delete. We describe each data structure with a focus on these three methods.

Concurrent Ordered Map

The implementation of the concurrent ordered map for persistent memory (pmem::obj::concurrent_map) is based on a concurrent skip list data structure. Intel TBB supplies tbb::concurrent_map, which is designed for volatile memory that we use as a baseline for a port to persistent memory. The concurrent skip list data structure can be implemented as a lock-free algorithm. But Intel chose a provably correct

---

[1] Intel Threading Building Blocks library (https://github.com/intel/tbb).

[2] Michael Voss, Rafael Asenjo, James Reinders. C++ Parallel Programming with Threading Building Blocks; Apress, 2019; ISBN-13 (electronic): 978-1-4842-4398-5; https://www.apress.com/gp/book/9781484243978.

scalable concurrent skip list[3] implementation with fine-grain locking distinguished by a combination of simplicity and scalability. Figure 14-3 demonstrates the basic idea of the skip list data structure. It is a multilayered linked list-like data structure where the bottom layer is an ordered linked list. Each higher layer acts as an "express lane" for the following lists and allows it to skip elements during lookup operations. An element in layer i appears in layer i+1 with some fixed probability p (in our implementation p = 1/2). That is, the frequency of nodes of a particular height decreases exponentially with the height. Such properties allow it to achieve O(log n) average time complexity for lookup, insert, and delete operations. O(log n) means the running time grows at most proportional to *"log n"*. You can learn more about *Big O notation* on Wikipedia at `https://en.wikipedia.org/wiki/Big_O_notation`

For the implementation of `pmem::obj::concurrent_map`, the `find` and `insert` operations are thread-safe and can be called concurrently with other `find` and `insert` operations without requiring additional synchronizations.

Find Operation

Because the `find` operation is non-modifying, it does not have to deal with data consistency issues. The `lookup` operation for the target element always begins from the topmost layer. The algorithm proceeds horizontally until the next element is greater or equal to the target. Then it drops down vertically to the next lower list if it cannot proceed on the current level. Figure 14-3 illustrates how the `find` operation works for the element with key=9. The search starts from the highest level and immediately goes from dummy head node to the node with key=4, skipping nodes with keys 1, 2, 3. On the node with key=4, the search is dropped two layers down and goes to the node with key=8. Then it drops one more layer down and proceeds to the desired node with key=9.

---

[3]M. Herlihy, Y. Lev, V. Luchangco, N. Shavit. A provably correct scalable concurrent skip list. In OPODIS '06: Proceedings of the 10th International Conference On Principles Of Distributed Systems, 2006; `https://www.cs.tau.ac.il/~shanir/nir-pubs-web/Papers/OPODIS2006-BA. pdf`.

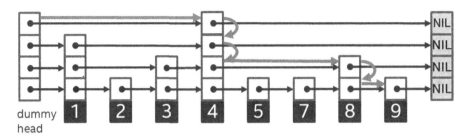

Figure 14-3. Finding key=9 in the skip list data structure

The find operation is wait-free. That is, every find operation is bound only by the number of steps the algorithm takes. And a thread is guaranteed to complete the operation regardless of the activity of other threads. The implementation of pmem::obj::concurrent_map uses atomic load-with-acquire memory semantics when reading pointers to the next node.

Insert Operation

The insert operation, shown in Figure 14-4, employs fine-grained locking schema for thread-safety and consists of the following basic steps to insert a new node with key=7 into the list:

1. Allocate the new node with randomly generated height.

2. Find a position to insert the new node. We must find the predecessor and successor nodes on each level.

3. Acquire locks for each predecessor node and check that the successor nodes have not been changed. If successor nodes have changed, the algorithm returns to step 2.

4. Insert the new node to all layers starting from the bottom one. Since the *find* operation is lock-free, we must update pointers on each level atomically using store-with-release memory semantics.

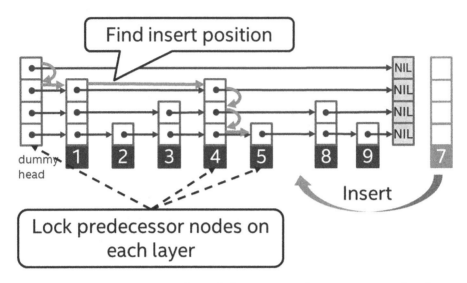

Figure 14-4. *Inserting a new node with key=7 into the concurrent skip list*

The algorithm described earlier is thread-safe, but it is not enough to be fault tolerant on persistent memory. There is a possible persistent memory leak if a program unexpectedly terminates between the first and fourth steps of our algorithm.

The implementation of `pmem::obj::concurrent_map` does not use transactions to support data consistency because transactions do not support isolation and by not using transactions, it can achieve better performance. For this linked list data structure, data consistency is maintained because a newly allocated node is always reachable (to avoid persistent memory leak) and the linked list data structure is always valid. To support these two properties, persistent thread-local storage is used, which is a member of the concurrent skip list data structure. Persistent thread-local storage guarantees that each thread has its own location in persistent memory to assign the result of persistent memory allocation for the new node.

Figure 14-5 illustrates the approach of this fault-tolerant `insert` algorithm. When a thread allocates a new node, the pointer to that node is kept in persistent thread-local storage, and the node is reachable through this persistent thread-local storage. Then the algorithm inserts the new node to the skip list by linking it to all layers using the thread-safe algorithm described earlier. Finally, the pointer in the persistent thread-local storage is removed because the new node is reachable now via skip list itself. In case of failure, a special function traverses all nonzero pointers in persistent thread-local storage and completes the `insert` operation.

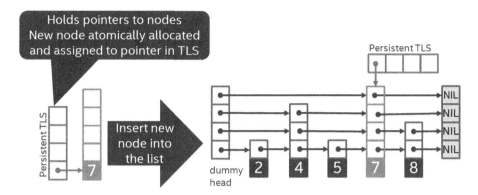

Figure 14-5. *Fault-tolerant insert operation using persistent thread-local storage*

Erase Operation

The implementation of the erase operation for pmem::obj::concurrent_map is not thread-safe. This method cannot be called concurrently with other methods of the concurrent ordered map because this is a memory reclamation problem that is hard to solve in C++ without a garbage collector. There is a way to logically extract a node from a skip list in a thread-safe manner, but it is not trivial to detect when it is safe to delete the removed node because other threads may still have access to the node. There are possible solutions, such as hazard pointers, but these can impact the performance of the find and insert operations.

Concurrent Hash Map

The concurrent hash map designed for persistent memory is based on tbb::concurrent_hash_map that exists in the Intel TBB. The implementation is based on a concurrent hash table algorithm where elements assigned to buckets based on a hash code are calculated from a key. In addition to concurrent find, insert, and erase operations, the algorithm employs concurrent resizing and on-demand per-bucket rehashing.[4]

Figure 14-6 illustrates the basic idea of the concurrent hash table. The hash table consists of an array of buckets, and each bucket consists of a list of nodes and a read-write lock to control concurrent access by multiple threads.

---

[4]Anton Malakhov. Per-bucket concurrent rehashing algorithms, 2015, arXiv:1509.02235v1; https://arxiv.org/ftp/arxiv/papers/1509/1509.02235.pdf.

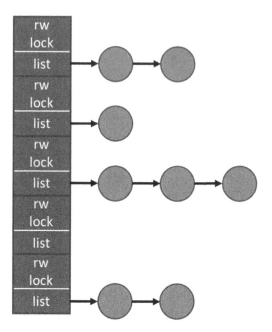

Figure 14-6. *The concurrent hash map data structure*

Find Operation

The find operation is a read-only event that does not change the hash map state. Therefore, data consistency is maintained while performing a find request. The find operation works by first calculating the hash value for a target key and acquires read lock for the corresponding bucket. The read lock guarantees that there is no concurrent modifications to the bucket while we are reading it. Inside the bucket, the find operation performs a linear search through the list of nodes.

Insert Operation

The insert method of the concurrent hash map uses the same technique to support data consistency as the concurrent skip list data structure. The operation consists of the following steps:

1. Allocate the new node, and assign a pointer to the new node to persistent thread-local storage.

2. Calculate the hash value of the new node, and find the corresponding bucket.

3. Acquire the write lock to the bucket.

4. Insert the new node to the bucket by linking it to the list of nodes. Because only one pointer has to be updated, a transaction is not needed. Because only one pointer is updated, a transaction is not required.

Erase Operation

Although the `erase` operation is similar to an `insert` (the opposite action), its implementation is even simpler than the `insert`. The `erase` implementation acquires the write lock for the required bucket and, using a transaction, removes the corresponding node from the list of nodes within that bucket.

Summary

Although building an application for persistent memory is a challenging task, it is more difficult when you need to create a multithreaded application for persistent memory. You need to handle data consistency in a multithreaded environment when multiple threads can update the same data in persistent memory.

If you develop concurrent applications, we encourage you to use existing libraries that provide concurrent data structures designed to store data in persistent memory. You should develop custom algorithms only if the generic ones do not fit your needs. See the implementations of concurrent `cmap` and `csmap` engines in `pmemkv`, described in Chapter 9, which are implemented using `pmem::obj::concurrent_hash_map` and `pmem::obj::concurrent_map`, respectively.

If you need to develop a custom multithreaded algorithm, be aware of the limitation PMDK transactions have for concurrent execution. This chapter shows that transactions do not automatically provide isolation out of the box. Changes made inside one transaction become visible to other concurrent transactions before they are committed. You will need to implement additional synchronization if it is required by an algorithm. We also explain that atomic operations cannot be used inside a transaction while building lock-free algorithms without transactions. This is a very complicated task if your platform does not support eADR.

## CHAPTER 15

Profiling and Performance

Introduction

This chapter first discusses the general concepts for analyzing memory and storage performance and how to identify opportunities for using persistent memory for both high-performance persistent storage and high-capacity volatile memory. We then describe the tools and techniques that can help you optimize your code to achieve the best performance.

Performance analysis requires tools to collect specific data and metrics about application, system, and hardware performance. In this chapter, we describe how to collect this data using Intel VTune Profiler. Many other data collection options are available; the techniques we describe are relevant regardless of how the data is collected.

Performance Analysis Concepts

Most concepts for performance analysis of persistent memory are similar to those already established for performance analysis of shared memory programs or storage bottlenecks. This section outlines several important performance considerations you should understand to profile and optimize persistent memory performance and defines the terms and situations we use in this chapter.

Compute-Bound vs. Memory-Bound

Performance optimization largely involves identifying the current performance bottleneck and improving it. The performance of compute-bound workloads is generally limited by the number of instructions the CPU can process per cycle. For example, an application doing a large number of calculations on very compact data without many dependencies is usually *compute-bound*. This type of workload would run faster if the CPU were faster. Compute-bound applications usually have high CPU utilization, close to 100%.

295

© The Author(s) 2020
S. Scargall, *Programming Persistent Memory*, https://doi.org/10.1007/978-1-4842-4932-1_15

In contrast, the performance of *memory-bound* workloads is generally limited by the memory subsystem and the long latencies of fetching data from caches and system memory. An example is an application that randomly accesses data from data structures in DRAM. In this case, adding more compute resources would not improve such an application. Adding persistent memory to improve performance is usually an option for memory-bound workloads as opposed to compute-bound workloads. Memory-bound workloads usually have lower CPU utilization than compute-bound workloads, exhibit CPU stalls due to memory transfers, and have high memory bandwidth.

Memory Latency vs. Memory Capacity

This concept is essential when discussing persistent memory. For this discussion, we assume that DRAM access latencies are lower than persistent memory and that the persistent memory capacity within the system is larger than DRAM. Workloads bound by memory capacity can benefit from adding persistent memory in a volatile mode, while workloads that are bound by memory latency are less likely to benefit.

Read vs. Write Performance

While each persistent memory technology is unique, it is important to understand that there is usually a difference in the performance of reads (loads) vs. writes (stores). Different media types exhibit varying degrees of asymmetric read-write performance characteristics, where reads are generally much faster than writes. Therefore, understanding the mix of loads and stores in an application workload is important for understanding and optimizing performance.

Memory Access Patterns

A memory access pattern is the pattern with which a system or application reads and writes to or from the memory. Memory hardware usually relies on temporal locality (accessing recently used data) and spatial locality (accessing contiguous memory addresses) for best performance. This is often achieved through some structure of fast internal caches and intelligent prefetchers. The access pattern and level of locality can drastically affect cache performance and can also have implications on parallelism and distributions of workloads within shared memory systems. Cache coherency can also affect multiprocessor performance, which means that certain memory access patterns place a ceiling on parallelism. Many well-defined memory access patterns exist, including but not limited to sequential, strided, linear, and random.

It is much easier to measure, control, and optimize memory accesses on systems that run only one application. In the cloud and virtualized environments, applications within the guests can be running any type of application and workload, including web servers, databases, or an application server. This makes it much harder to ensure memory accesses are fully optimized for the hardware as the access patterns are essentially random.

I/O Storage Bound Workloads

A program is I/O bound if it would go faster if the I/O subsystem were faster. We are primarily interested in the block-based disk I/O subsystem here, but it could also include other subsystems such as the network. An I/O bound state is undesirable because it means that the CPU must stall its operation while waiting for data to be loaded or unloaded from main memory or storage. Depending on where the data is and the latency of the storage device, this can invoke a voluntary context switching of the current application thread with another. A voluntary context switch occurs when a thread blocks because it requires a resource that is not immediately available or takes a long time to respond. With faster computation speed being the primary goal of each successive computer generation, there is a strong imperative to avoid I/O bound states. Eliminating them can often yield a more economic improvement in performance than upgrading the CPU or memory.

Determining the Suitability of Workloads for Persistent Memory

Persistent memory technologies may not solve every workload performance problem. You should understand the workload and platform on which it is currently running when considering persistent memory. As a simple example, consider a compute-intensive workload that relies heavily on floating-point arithmetic. The performance of this application is likely limited by the floating-point unit in the CPU and not any part of the memory subsystem. In that case, adding persistent memory to the platform will likely have little impact on this application's performance. Now consider an application that requires extensive reading and writing from disk. It is likely that the disk accesses are the bottleneck for this application and adding a faster storage solution, like persistent memory, could improve performance.

These are trivial examples, and applications will have widely different behaviors along this spectrum. Understanding what behaviors to look for and how to measure them is an important step to using persistent memory. This section presents the important characteristics to identify and determine if an application is a good fit for persistent memory. We look at applications that require in-memory persistence, applications that can use persistent memory in a volatile manner, and applications that can use both.

Volatile Use Cases

Chapter 10 described several libraries and use cases where applications can take advantage of the performance and capacity of persistent memory to store non-volatile data. For volatile use cases, persistent memory will act as an additional memory tier for the platform. It may be transparent to the application, such as using Memory Mode supported by Intel Optane DC persistent memory, or applications can make code changes to perform volatile memory allocations using libraries such as libmemkind. In both cases, memory-capacity bound workloads will benefit from adding persistent memory to the platform. Application performance can dramatically improve if its working dataset can fit into memory and avoid paging to disk.

Identifying Workloads That Are Memory-Capacity Bound

To determine if a workload is memory-capacity bound, you must determine the "memory footprint" of the application. The memory footprint is the high watermark of memory concurrently allocated during the application's life cycle. Since physical memory is a finite resource, you should consider the fact that the operating system and other processes also consume memory. If the footprint of the operating system and all memory consumers on the system are approaching or exceeding the available DRAM capacity on the platform, you can assume that the application would benefit from additional memory because it cannot fit all its data in DRAM. Many tools and techniques can be used to determine memory footprint. VTune Profiler includes two different ways to find this information: *Memory Consumption* analysis or *Platform Profiler* analysis. VTune Profiler is a free download for Linux and Windows, available from `https://software.intel.com/en-us/vtune`.

The Memory Consumption analysis within VTune Profiler tracks all memory allocations made by the application. Figure 15-1 shows a VTune Profiler bottom-up report representing memory consumption of the profiled application over time. The highest value on the y-axis in the Memory Consumption timeline indicates the application footprint is approximately 1GiB.

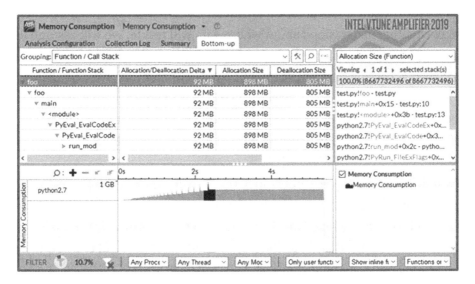

Figure 15-1. *The VTune Profiler bottom-up analysis showing memory consumption with time and the associated allocating call stacks*

The Memory Utilization graph in the Platform Profiler analysis shown in Figure 15-2 measures the memory footprint using operating system statistics and produces a timeline graph as a percentage of the total available memory.

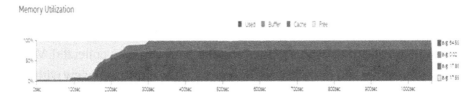

Figure 15-2. *The VTune Platform Profiler Memory Utilization graph as a percentage of total system memory*

The results in Figure 15-2 were taken from a different application than Figure 15-1. This graph shows very high memory consumption, which implies this workload would be a good candidate for adding more memory to the system. If your persistent memory hardware has variable modes, like the Memory and App Direct modes on Intel Optane DC persistent memory, you will need some more information to determine which mode to use first. The next important information is the hot working set size.

Identifying the Hot Working Set Size of a Workload

Persistent memory usually has different characteristics than DRAM; therefore, you should make intelligent decisions about where data will reside. We will assume that accessing data from persistent memory has higher latency than DRAM. Given the choice between accessing data in DRAM and persistent memory, we would always choose DRAM for performance. However, the premise of adding persistent memory in a volatile configuration assumes there is not enough DRAM to fit all the data. You need to understand how your workload accesses data to make choices about persistent memory configuration.

The working set size (WSS) is how much memory an application needs to keep working. For example, if an application has 50GiB of main memory allocated and page mapped, but it is only accessing 20MiB each second to perform its job, we can say that the working set size is 50GiB and the "hot" data is 20MiB. It is useful to know this for capacity planning and scalability analysis. The "hot working set" is the set of objects accessed frequently by an application, and the "hot working set size" is the total size of those objects allocated at any given time.

Determining the size of the working set and hot working set is not as straightforward as determining memory footprint. Most applications will have a wide range of objects with varying degrees of "hotness," and there will not be a clear line delineating which objects are hot and which are not. You must interpret this information and determine the hot working set size.

VTune Profiler has a Memory Access analysis feature that can help determine the hot and working set sizes of an application (select the "Analyze dynamic memory objects" option before data collection begins). Once enough data has been collected, VTune Profiler will process the data and produce a report. In the bottom-up view within the GUI, a grid lists each memory object that was allocated by the application.

Figure 15-3 shows the results of a Memory Access analysis of an application. It shows the memory size in parenthesis and the number of loads and stores that accessed it. The report does not include an indication of what was concurrently allocated.

Grouping: **Memory Object / Function / Call Stack**			
Memory Object / Function / Call Stack	Loads ▾	Stores	LLC Miss Count »
▶ matrix.c:116 (128 MB)	161,578,247,202	0	0
▶ matrix.c:121 (128 MB)	15,043,951,305	0	0
▶ matrix.c:126 (128 MB)	2,196,965,907	70,028,400,789	2,250,135
▶ [vmlinux]	117,903,537	65,701,971	0

Figure 15-3. *Objects accessed by the application during a Memory Access analysis data collection*

The report identifies the objects with the most accesses (loads and stores). The sum of the sizes of these objects is the working set size – the values are in parentheses. You decide where to draw the line for what is and is not part of the hot working set.

Depending on the workload, there may not be an easy way to determine the hot working set size, other than developer knowledge of the application. Having a rough estimate is important for deciding whether to start with Memory Mode or App Direct mode.

Use Cases Requiring Persistence

Use cases that take advantage of persistent memory for persistence, as opposed to the volatile use cases previously described, are generally replacing slower storage devices with persistent memory. Determining the suitability of a workload for this use case is straightforward. If application performance is limited by storage accesses (disks, SSDs, etc.), then using a faster storage solution like persistent memory could help. There are several ways to identify storage bottlenecks in an application. Open source tools like dstat or iostat give a high-level overview of disk activity, and tools such as VTune Profiler provide a more detailed analysis.

nvme0n1

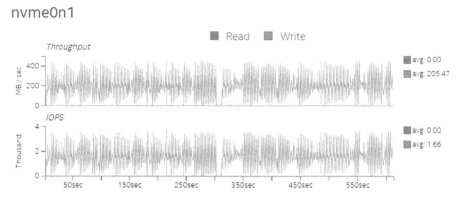

Figure 15-4. *Disk throughput and IOPS graphs from VTune Profiler's Platform Profiler*

Figure 15-4 shows throughput and IOPS numbers of an NVMe drive collected using Platform Profiler. This example uses a non-volatile disk for extensive storage, as indicated by the throughput and IOPS graphs. Applications like this may benefit from faster storage like persistent memory. Another important metric to identify storage bottlenecks is I/O Wait time. The Platform Profiler analysis can also provide this metric and display how it is affecting CPU Utilization over time, as seen in Figure 15-5.

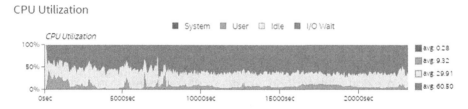

Figure 15-5. *I/O Wait time from VTune Profiler's Platform Profiler*

Performance Analysis of Workloads Using Persistent Memory

Optimizing a workload on a system with persistent memory follows the principles similar to those of optimizing a workload performance on a DRAM-only system. The additional factors to keep in mind are:

- The writes to persistent memory may impact performance more than the reads.

- Applications can allocate objects on DRAM or persistent memory. If done indiscriminately, this can negatively impact performance.

- In Memory Mode (specific to Intel Optane DC persistent memory), users have the option of varying the near-memory cache size (DRAM size) to improve workload performance.

Keeping these additional factors in mind, the approach to workload performance optimization will follow the same process of characterizing the workload, choosing the correct memory configuration, and optimizing the code for maximum performance.

Characterizing the Workload

The performance of a workload on a persistent memory system depends on a combination of the workload characteristics and the underlying hardware. The key metrics to understand the workload characteristics are:

- Persistent memory bandwidth

- Persistent memory read/write ratio

- Paging to and from traditional storage

- Working set size and footprint of the workload

- Nonuniform Memory Architecture (NUMA) characteristics

- Near-memory cache behavior in Memory Mode (specific to Intel Optane DC persistent memory)

Memory Bandwidth and Latency

Persistent memory, like DRAM, has limited bandwidth. When it becomes saturated, it can quickly bottleneck application performance. Bandwidth limits will vary depending on the platform. You can calculate the peak bandwidth of your platform using hardware specifications or a memory benchmarking application.

The Intel Memory Latency Checker (Intel MLC) is a free tool for Linux and Windows available from `https://software.intel.com/en-us/articles/intelr-memory-latency-checker`. Intel MLC can be used to measure bandwidth and latency of DRAM and persistent memory using a variety of tests:

- Measure idle memory latencies between each CPU socket

- Measure peak memory bandwidth requests with varying ratios of reads and writes

- Measure latencies at different bandwidth points

- Measure latencies for requests addressed to a specific memory controller from a specific core

- Measure cache latencies

- Measure b/w from a subset of the cores/sockets

- Measure b/w for different read/write ratios

- Measure latencies for random and sequential address patterns

- Measure latencies for different stride sizes

- Measure cache-to-cache data transfer latencies

VTune Profiler has a built-in kernel to measure peak bandwidth on a system. Once you know the peak bandwidth of the platform, you can then measure the persistent memory bandwidth of your workload. This will reveal whether persistent memory bandwidth is a bottleneck. Figure 15-6 shows an example of persistent memory read and write bandwidth of an application.

Figure 15-6. *Results from VTune Profiler persistent memory bandwidth measurements*

Persistent Memory Read-Write Ratio

As described in "Performance Analysis Concepts," the ratio of read and write traffic to the persistent memory plays a major role in the overall performance of a workload. If the ratio of persistent memory write bandwidth to read bandwidth is high, there is a good chance the persistent memory write latency is impacting performance. Using the Platform Profiler feature in VTune Profiler is one way to collect this information. Figure 15-7 shows the ratio of read traffic vs. all traffic to persistent memory. This number should be close to 1.0 for best performance.

Figure 15-7.   *Read traffic ratio from VTune Profiler's Platform Profiler analysis*

Working Set Size and Memory Footprint

As described in "Determining the Suitability of Workloads for Persistent Memory," the working set size and memory footprint of the application are important characteristics to understand once a workload is running on a system with persistent memory. Metrics can be collected using the tools and processes previously described.

Non-Uniform Memory Architecture (NUMA) Behavior

Multi-socket platforms typically have persistent memory attached to each socket. Accesses to persistent memory from a thread on one socket to another will incur longer latencies. These "remote" accesses are some of the NUMA behaviors that can impact performance. Multiple metrics can be collected to determine how much NUMA activity is occurring in a workload. On Intel platforms, data moves between sockets through the socket interconnect called the QuickPath Interconnect (QPI) or Ultra Path Interconnect (UPI). High interconnect bandwidth may indicate NUMA-related performance issues. In addition to interconnect bandwidth, some hardware provides counters to track local and remote accesses to persistent memory.

Understanding the NUMA behavior of your workload is another important step in understanding performance optimization. Figure 15-8 shows UPI bandwidth collected with VTune Profiler.

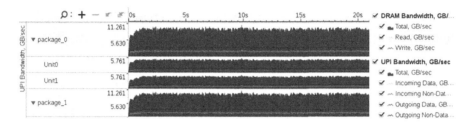

Figure 15-8. *UPI traffic ratio from VTune Profiler*

The Platform Profiler feature in VTune Profiler can collect metrics specific to persistent memory.

Tuning the Hardware

The memory configuration of a system is a significant factor in determining the system's performance. The workload performance depends on a combination of workload characteristics and the memory configuration. There is no single configuration that provides the best value for all workloads. These factors make it important to tune the hardware with respect to workload characteristics and get the maximum value out of the system.

Addressable Memory Capacity

The combined capacity of DRAM and persistent memory determines the total addressable memory available on the system. You should tune the size of persistent memory to accommodate the workload's footprint.

The capacity of DRAM available on the system should be large enough to accommodate the workload's hot working set size. A large amount of volatile traffic going to persistent memory while DRAM is fully utilized is a good indicator that the workload can benefit from additional DRAM size.

Bandwidth Requirements

The maximum available persistent memory bandwidth depends on the number of channels populated with a persistent memory module. A fully populated system works well for a workload with a high bandwidth requirement. Partially populated systems can be used for workloads that are not as memory latency sensitive. Refer to the server documentation for population guidelines.

BIOS Options

With the introduction of persistent memory into server platforms, many features and options have been added to the BIOS that provide additional tuning capabilities. The options and features available within the BIOS vary for each server vendor and persistent memory product. Refer to the server BIOS documentation for all the options available; most share common options, including:

- Ability to change power levels to balance power consumption and performance. More power delivered to persistent memory can increase performance

- Enable or disable persistent memory–specific features

- Tune latency or bandwidth characteristics of persistent memory

Optimizing the Software for Persistent Memory

There are many ways to optimize applications to use persistent memory efficiently and effectively. Each application will benefit in different ways and will need to have code modified accordingly. This section describes some of the optimization methods.

Guided Data Placement

Guided data placement is the most common avenue for optimizing volatile workloads on a persistent memory system. Application developers can choose to allocate a data structure or object in DRAM or persistent memory. It is important to choose accurately because allocating incorrectly could impact application performance. This allocation is usually handled via specific APIs, for example, the allocation APIs available in the Persistent Memory Development Kit (PMDK) and memkind library.

Depending on your familiarity with the code and how it works with production workloads, knowing which data structures and objects to store in the different memory/storage tiers may be simple. Should those data structures and objects be volatile or persisted? To help with searching for potential candidates, tools such as VTune Profiler can identify objects with the most last-level cache (LLC) misses. The intent is to identify what data structures and objects the application uses most frequently and ensure they are placed in the fastest media appropriate to their access patterns. For example, an object that is written once but read many times is best placed in DRAM. An object that is updated frequently that needs to be persisted should probably be moved to persistent memory rather than traditional storage devices.

You must also be mindful of memory-capacity constraints. Tools such as VTune Profiler can help determine approximately how many hot objects will fit into the available DRAM. For the remaining objects that have fewer LLC misses or that are too large to allocate from DRAM, you can put them in persistent memory. These steps will ensure that your most accessed objects have the fastest path to the CPU (allocated in DRAM), while the infrequently accessed objects will take advantage of the additional persistent memory (as opposed to sitting out on a much slower storage devices).

Another consideration for optimizations is the load/store ratio for object accesses. If your persistent memory hardware characteristics are such that load/read operations are much faster than stores/writes, this should be taken into account. Objects with a high load/store ratio should benefit from living in persistent memory.

There is no hard rule for what constitutes a frequent vs. infrequently accessed object. Although behaviors are application dependent, these guidelines give a starting point for choosing how to allocate objects in persistent memory. After completing this process, start profiling and tuning the application to further improve the performance with persistent memory.

Memory Access Optimization

The common techniques for optimizing cache performance on DRAM-only platforms also apply to persistent memory platforms. Concepts like cache-miss penalties and spatial/temporal data locality are important for performance. Many tools can collect performance data for caches and memory. VTune Profiler has predefined metrics for each level of the memory hierarchy, including Intel Optane DC persistent memory shown in Figure 15-9.

⌄ Memory Bound ⑦:	**69.6%** ⚑	of Pipeline Slots
L1 Bound ⑦:	**12.5%** ⚑	of Clockticks
L2 Bound ⑦:	**0.4%**	of Clockticks
L3 Bound ⑦:	**2.9%**	of Clockticks
⌃ DRAM Bound ⑦:	**0.0%** ⚑	of Clockticks
⌄ Persistent Memory Bound ⑦:	**49.0%**	of Clockticks
Persistent Memory Bandwidth Bound ⑦:	**0.0%**	**of Elapsed Time**
Local Persistent Memory ⑦:	**100.0%**	of Clockticks
Remote Persistent Memory ⑦:	**0.0%**	of Clockticks

Figure 15-9. *VTune Profiler memory analysis of a workload showing a breakdown of CPU cache, DRAM, and persistent memory accesses*

These performance metrics help to determine if memory is the bottleneck in your application, and if so, which level of the memory hierarchy is the most impactful. Many tools can pinpoint source code locations and memory objects responsible for the bottleneck. If persistent memory is the bottleneck, review the "Guided Data Placement" section to ensure that persistent memory is being used efficiently. Performance optimizations like cache blocking, software prefetching, and improved memory access patterns may also help relieve bottlenecks in the memory hierarchy. You must determine how to refactor the software to more efficiently use memory, and metrics like these can point you in the right direction.

NUMA Optimizations

NUMA-related performance issues were described in the "Characterizing the Workload" section; we discuss NUMA in more detail in Chapter 19. If you identify performance issues related to NUMA memory accesses, two things should be considered: data allocation vs. first access, and thread migration.

Data Allocation vs. First Access

Data allocation is the process of allocating or reserving some amount of virtual address space for an object. The virtual address space for a process is the set of virtual memory addresses that it can use. The address space for each process is private and cannot be accessed by other processes unless it is shared. A virtual address does not represent the actual physical location of an object in memory. Instead, the system maintains a multilayered page table, which is an internal data structure used to translate virtual addresses into their corresponding physical addresses. Each time an application thread references an address, the system translates the virtual address to a physical address. The physical address points to memory physically connected to a CPU. Chapter 19 describes exactly how this operation works and shows why high-capacity memory systems can benefit from using large or huge pages provided by the operating system.

A common practice in software is to have most of the data allocations done when the application starts. Operating systems try to allocate memory associated with the CPU on which the thread executes. The operating system scheduler then tries to always schedule the thread on a CPU that it last ran in the hopes that the data still remains in one of the CPU caches. On a multi-socket system, this may result in all the objects being allocated in the memory of a single socket, which can create NUMA performance issues. Accessing data on a remote CPU incurs a latency performance penalty.

Some applications delay reserving memory until the data is accessed for the first time. This can alleviate some NUMA issues. It is important to understand how your workload allocates data to understand the NUMA performance.

Thread Migration

Thread migration, which is the movement of software threads across sockets by the operating system scheduler, is the most common cause of NUMA issues. Once objects are allocated in memory, accessing them from another physical CPU from which they were originally allocated incurs a latency penalty. Even though you may allocate your data on a socket where the accessing thread is currently running, unless you have specific affinity bindings or other safeguards, the thread may move to any other core or socket in the future. You can track thread migration by identifying which cores threads are running on and which sockets those cores belong to. Figure 15-10 shows an example of this analysis from VTune Profiler.

Thread / Package / Core / Function / Call Stack	CPU Time ▼	Instructions Retired	Microarchitecture Usage	TID
▶ matrix.gcc (TID: 455334)	3.844s	1,255,500,000	9.6%	455334
▼ matrix.gcc (TID: 455336)	3.834s	1,215,000,000	11.8%	455336
▼ package_1	3.528s	1,053,000,000	9.8%	455336
▶ core_17	2.551s	729,000,000	7.1%	455336
▶ core_1	0.386s	121,500,000	16.9%	455336
▶ core_7	0.286s	81,000,000	28.0%	455336
▶ core_5	0.010s	0	0.0%	455336
▶ package_0	0.306s	162,000,000	30.1%	455336

Figure 15-10. *VTune Profiler identifying thread migration across cores and sockets (packages)*

Use this information to determine whether thread migration is correlated with NUMA accesses to remote DRAM or persistent memory.

Large and Huge Pages

The default memory page size in most operating systems is 4 kilobytes (KiB). Operating systems provide many different page sizes for different application workloads and requirements. In Linux, a *Large Page* is 2 megabytes (MiB), and a *Huge Page* is 1 gigabyte (GiB). The larger page sizes can be beneficial to workload performance on persistent memory in certain scenarios.

For applications with a large addressable memory requirement, the size of the page table being maintained by the operating system for virtual to physical address translation grows significantly larger in size. The translation lookaside buffer (TLB) is a small cache to make virtual-to-physical address translations faster. The efficiency of TLB goes down when the number of page entries increases in the page table. Chapter 19 describes this in more detail.

Persistent memory systems that are meant for applications with a large memory requirement will likely encounter the problem of large page tables and inefficient TLB usage. Using large page sizes in this scenario helps reduce the number of entries in the page table. The main trade-offs when using large page sizes is a higher overhead for each allocation and memory fragmentation. You must be aware of the application behavior before using large pages on persistent memory. An application doing frequent allocation/deallocation may not be a good fit for large page optimization. The memory fragmentation issue is somewhat abated by the large address space available on the persistent memory systems.

Summary

Profiling and performance optimization techniques for persistent memory systems are similar to those techniques used on systems without persistent memory. This chapter outlined some important concepts for understanding performance. It also provides guidance for characterizing an existing application without persistent memory and understanding whether it is suitable for persistent memory. Finally, it presents important metrics for performance analysis and tuning of applications running on persistent memory platforms, including some examples of how to collect the data using the VTune Profiler tool.

Performance profiling and optimization are an iterative process that only ends when you determine that the investment required for the next improvement is too high for the benefit that will be returned. Use the concepts introduced in this chapter to understand how your workloads can benefit from persistent memory, and use some of the optimization techniques we discussed to tune for this type of platform.

CHAPTER 16

PMDK Internals: Important Algorithms and Data Structures

Chapters 5 through 10 describe most of the libraries contained within the Persistent Memory Development Kit (PMDK) and how to use them.

This chapter introduces the fundamental algorithms and data structures on which libpmemobj is built. After we first describe the overall architecture of the library, we discuss the individual components and the interaction between them that makes libpmemobj a cohesive system.

A Pool of Persistent Memory: High-Level Architecture Overview

Figure 16-1 shows that libpmemobj comprises many isolated components that build on top of each other to provide a transactional object store.

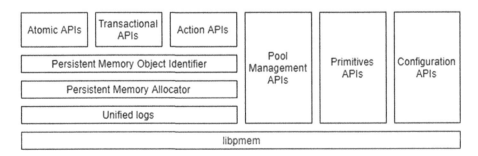

Figure 16-1. *The modules of the libpmemobj architecture*

313

© The Author(s) 2020
S. Scargall, *Programming Persistent Memory*, https://doi.org/10.1007/978-1-4842-4932-1_16

Everything is built on top of libpmem and its persistence primitives that the library uses to transfer data to persistent memory and persist it. Those primitives are also exposed through libpmemobj-specific APIs to applications that wish to perform low-level operations on persistent memory, such as manual cache flushing. These APIs are exposed so the high-level library can instrument, intercept, and augment all stores to persistent memory. This is useful for the instrumentation of runtime analysis tools such as Valgrind pmemcheck, described in Chapter 12. More importantly, these functions are interception points for data replication, both local and remote.

Replication is implemented in a way that ensures all data written prior to calling drain will be safely stored in the replica as configured. A drain operation is a barrier that waits for hardware buffers to complete their flush operation to ensure all writes have reached the media. This works by initiating a write to the replica when a memory copy or a flush is performed and then waits for those writes to finish in the drain call. This mechanism guarantees the same behavior and ordering semantics for replicated and non-replicated pools.

On top of persistence primitives provided by libpmem is an abstraction for fail-safe modification of transactional data called *unified logging*. The unified log is a single data structure and API for two different logging types used throughout libpmemobj to ensure fail-safety: transactions and atomic operations. This is one of the most crucial, performance-sensitive modules in the library because it is the hot code path of almost every API. The unified log is a hybrid DRAM and persistent memory data structure accessed through a runtime context that organizes all memory operations that need to be performed within a single fail-safe atomic transaction and allows for critical performance optimizations.

The persistent memory allocator operates in the unified log context of either a transaction or a single atomic operation. This is the largest and most complex module in libpmemobj and is used to manage the potentially large amounts of persistent memory associated with the memory pool.

Each object stored in a persistent memory pool is represented by an object handle of type PMEMoid (persistent memory object identifier). In practice, such a handle is a unique object identifier (OID) of global scope, which means that two objects from different pools will never have the same OID. An OID cannot be used as a direct pointer to an object. Each time the program attempts to read or write object data, it must obtain the current memory address of the object by converting its OID into a pointer. In contrast to the memory address, the OID value for a given object does not change during the life of an object, except for a realloc(), and remains valid after closing and reopening the pool. For this reason, if an object contains a reference to another persistent object, for example, to build a linked data structure, the reference must be an OID and not a memory address.

The atomic and transactional APIs are built using a combination of the persistent memory allocator and unified logs. The simplest public interface is the atomic API which runs a single allocator operation in a unified log context. That log context is not exposed externally and is created, initialized, and destroyed within a single function call.

The most general-purpose interface is the transactional API, which is based on a combination of undo logging for snapshots and redo logging for memory allocation and deallocation. This API has ACID (atomicity, consistency, isolation, durability) properties, and it is a relatively thin layer that combines the utility of unified logs and the persistent memory allocator.

For specific transactional use cases that need low-level access to the persistent memory allocator, there is an "action" API. The action API is essentially a pass-through to the raw memory allocator interface, alongside helpers for usability. This API can be leveraged to create low-overhead algorithms that issue fewer memory fences, as compared to general-purpose transactions, at the cost of ease of use.

All public interfaces produce and operate on PMEMoids as a replacement for pointers. This comes with space overhead because PMEMoids are 16 bytes. There is also a performance overhead for the translation to a normal pointer. The upside is that objects can be safely referenced between different instances of the application and even different persistent memory pools.

The pool management API opens, maps, and manages persistent memory resident files or devices. This is where the replication is configured, metadata and the heap are initialized, and all the runtime data is created. This is also where the crucial recovery of interrupted transactions happens. Once recovery is complete, all prior transactions are either committed or aborted, the persistent state is consistent, and the logs are clean and ready to be used again.

The Uncertainty of Memory Mapping: Persistent Memory Object Identifier

A key concept that is important for any persistent memory application is how to represent the relative position of an object within a pool of memory, and even beyond it. That is, how do you implement pointers? You could rely on normal pointers, which are relative to the beginning of the application's virtual address space, but that comes with many caveats. Using such pointers would be predicated on the pool of persistent memory always being located at the same place in the virtual address space of an application that maps it. This is difficult, if not impossible, to accomplish in a portable way on modern operating systems due to address space layout randomization (ASLR). Therefore, a general-purpose library for persistent memory programming must provide a specialized persistent pointer. Figure 16-2 shows a pointer from Object A to Object B. If the base address changes, the pointer no longer points to Object B.

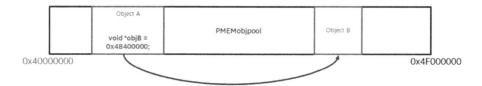

Figure 16-2. *Example of using a normal pointer in a persistent memory pool*

An implementation of a general-purpose relative persistent pointer should satisfy these two basic requirements:

1. The pointer must remain valid across application restarts.

2. The pointer should unambiguously identify a memory location in the presence of many persistent memory pools, even if not located in a pool from which it was originally derived.

In addition to the previous requirements, you should also consider some potential performance problems:

- Additional space overhead over a traditional pointer. This is important because large fat pointers would take up more space in memory and because fewer of these fat pointers would fit in a single CPU cache line. This potentially increases the cost of operations in pointer-chasing heavy data structures, such as those found in B-tree algorithms.

- The cost of translating persistent pointers to real pointers. Because dereferencing is an extremely common operation, this calculation must be as lightweight as possible and should involve as few instructions as possible. This is to ensure that persistent pointer usage is efficient and it doesn't generate too much code bloat during compilation.

- Preventing compiler optimizations through the dereferencing method. A complicated pointer translation might negatively impact the compiler's optimization passes. The translation method should ideally avoid operations that depend on an external state because that will prevent, among other things, auto-vectorization.

Satisfying the preceding requirements while maintaining low-overhead and C99 standard compliance is surprisingly difficult. We explored several options:

- The 8-byte offset pointer, relative to the beginning of the pool, was quickly ruled out because it did not satisfy the second requirement and needed a pool base pointer to be provided to the translation method.

- 8-byte self-relative pointers, where the value of the pointer is the offset between the object's location and the pointer's location. This is potentially the fastest implementation because the translation method can be implemented as `ptr + (*ptr)`. However, this does not satisfy the second basic requirement. Additionally, it would require a special assignment method because the value of the pointer to the same object would differ depending on the pointer's location.

- 8-byte offset pointers with embedded memory pool identifier, which allows the library to satisfy the second requirement. This is an augmentation of the first method that additionally stores the identifier in the unused part of the pointer value by taking advantage of the fact that the usable size of the virtual address space is smaller than the size of the pointer on most modern CPUs. The problem with this method, however, is that the number of bits for the pool identifier is relatively small (16 bits on modern CPUs) and might shrink with future hardware.

- 16-byte fat offset pointer with pool identifier. This is the most obvious solution, which is similar to the one earlier but has 8-byte offset pointers and 8-byte pool identifiers. Fat pointers provide the best utility, at the cost of space overhead and some runtime performance.

libpmemobj uses the most generic approach of the 16-byte offset pointer. This allows you to make your own choice since all other pointer types can be directly derived from it. libpmemobj bindings for more expressive languages than C99, such as C++, can also provide different types of pointers with different trade-offs.

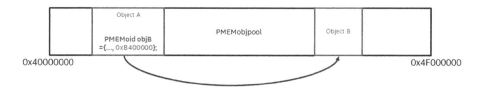

Figure 16-3. *Example of using a PMEMoid in a persistent memory pool*

Figure 16-3 shows the translation method used to convert a libpmemobj persistent pointer, PMEMoid, into a valid C pointer. In principle, this approach is very simple. We look up the base address of the pool through the pool identifier and then add the object offset to it. The method itself is static inline and defined in the public header file for libpmemobj to avoid a function call on every deference. The problem is the lookup method, which, for an application linked with a dynamic library, means a call to a different compilation unit, and that might be costly for a very commonly performed operation. To resolve this problem, the translation method has a per-thread cache of the last base address, which removes the necessity of calling the lookup with each dereferencing for the common case where persistent pointers from the same pool are accessed close together.

The pool lookup method itself is implemented using a radix tree that stores identifier-address pairs. This tree has a lock-free read operation, which is necessary because each non-cached pointer translation would otherwise have to acquire a lock to be thread-safe, and that would have a severe negative performance impact and could potentially serialize access to persistent memory.

Persistent Thread Local Storage: Using Lanes

Very early in the development of PMDK, we found that persistent memory programming closely resembles multithreaded programming because it requires restricting visibility of memory changes – either through locking or transactions – to other threads or instances of the program. But that is not the only similarity. The other similarity, which we discuss in this section, is how sometimes low-level code needs to store data that is unique to one thread of execution. In the persistent case, we often need to associate data with a transaction rather than a thread.

In `libpmemobj`, we need a way to create an association between an in-flight transaction and its persistent logs. It also requires a way to reconnect to those logs after an unplanned interruption. The solution is to use a data structure called a "lane," which is simply a persistent byte buffer that is also transaction local.

Lanes are limited in quantity, have a fixed size, and are located at the beginning of the pool. Each time a transaction starts, it chooses one of the lanes to operate from. Because there is a limited number of lanes, there is also a limited number of transactions that can run in parallel. For this reason, the size of the lane is relatively small, but the number of lanes is big enough as to be larger than a number of application threads that could feasibly run in parallel on current platforms and platforms coming in the foreseeable future.

The challenge of the lane mechanism is the selection algorithm, that is, which lane to choose for a specific transaction. It is a scheduler that assigns resources (lanes) to perform work (transactions).

The naive algorithm, which was implemented in the earliest versions of `libpmemobj`, simply picked the first available lane from the pool. This approach has a few problems. First, the implementation of what effectively amounts to a single LIFO (last in, first out) data structure of lanes requires a lot of synchronization on the front of the stack, regardless of whether it is implemented as a linked list or an array, and thus reducing performance. The second problem is false sharing of lane data. False sharing occurs when two or more threads operate on data that is being modified, causing CPU cache thrashing. And that is exactly what happens if multiple threads are continually fighting over the same number of lanes to start new transactions. The third problem is spreading the traffic across interleaved DIMMs. Interleaving is a technique that allows sequential traffic to take advantage of throughput of all of the DIMMs in the interleave set by spreading the physical memory across all available DIMMs. This is similar to striping (RAID0) across multiple disk drives. Depending on the size of the interleaved block, and the platform configuration, using naive lane allocation might continuously use the same physical DIMMs, lowering the overall performance.

To alleviate these problems, the lane scheduling algorithm in `libpmemobj` is more complex. Instead of using a LIFO data structure, it uses an array of 8-byte spinlocks, one for each lane. Each thread is initially assigned a primary lane number, which is assigned in such a way as to minimize false sharing of both lane data and the spinlock array.

The algorithm also tries to spread the lanes evenly across interleaved DIMMs. As long as there are fewer active threads than lanes, no thread will ever share a lane. When a thread attempts to start a transaction, it will try to acquire its primary lane spinlock, and if it is unsuccessful, it will try to acquire the next lane in the array.

The final lane scheduling algorithm decision took a considerable amount of research into various lane scheduling approaches. Compared to the naive implementation, the current implementation has vastly improved performance, especially in heavily multithreaded workloads.

Ensuring Power-Fail Atomicity: Redo and Undo Logging

The two fundamental concepts libpmemobj uses to ensure power-fail safety are redo and undo logging. Redo logs are used to ensure atomicity of memory allocations, while undo logs are used to implement transactional snapshots. Before we discuss the many different possible implementation approaches, this section describes the basic ideas.

Transaction Redo Logging

Redo logging is a method by which a group of memory modifications that need to be done atomically are stored in a log and deferred until all modifications in the group are persistently stored. Once completed, the log is marked as complete, and the memory modifications are processed (applied); the log can then be discarded. If the processing is interrupted before it finishes, the logging is repeated until successful. Figure 16-4 shows the four phases of transaction redo logging.

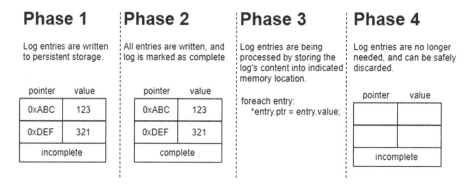

Figure 16-4. *The phases of a transaction redo log*

The benefit of this logging approach, in the context of persistent memory, is that all the log entries can be written and flushed to storage at once. An optimal implementation of redo logging uses only two synchronization barriers: once to mark the log as complete and once to discard it. The downside to this approach is that the memory modifications are not immediately visible, which makes for a more complicated programming model. Redo logging can sometimes be used alongside load/store instrumentation techniques which can redirect a memory operation to the logged location. However, this approach can be difficult to implement efficiently and is not well suited for a general-purpose library.

Transaction Undo Logging

Undo logging is a method by which each memory region of a group (undo transaction) that needs to be modified atomically is snapshotted into a log prior to the modification. Once all memory modifications are complete, the log is discarded. If the transaction is interrupted, the modifications in the log are rolled back to their original state. Figure 16-5 shows the three phases of the transaction undo logging.

Phase 1	**Phase 2**	**Phase 3a**	**Phase 3b**
Snapshot is written to the log.	Memory is modified in-place.	If the transaction was completed successfully, the log is discarded.	If the transaction was aborted, the log is processed, restoring the old values.

Phase 1:

pointer	old_value
0xABC	321

Phase 2: `*(entry.ptr) = 123;`

Phase 3a:

pointer	old_value

Phase 3b:
```
foreach entry:
    *entry.ptr = entry.old_value;
```

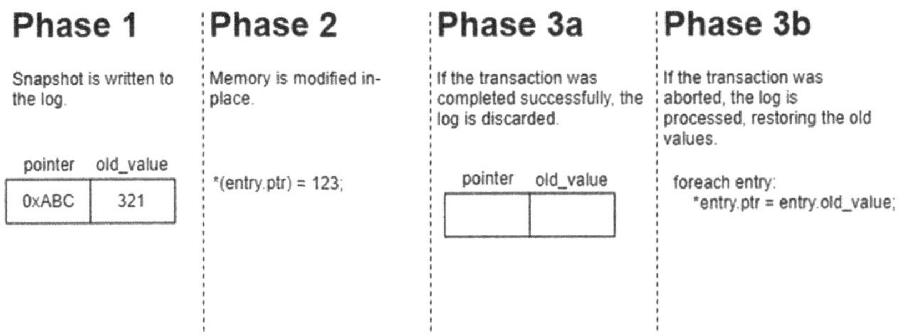

Figure 16-5. Phases of a transaction undo log

This type of log can have lower performance characteristics compared with the redo log approach because it requires a barrier for every snapshot that needs to be made, and the snapshotting itself must be fail-safe atomic, which presents its own challenges. An undo log benefit is that the changes are visible immediately, allowing for a natural programming model.

The important observation here is that redo and undo logging are complimentary. Use redo logging for performance-critical code and where deferred modifications are not a problem; use undo logging where ease of use is important. This observation led to the current design of libpmemobj where a single transaction takes advantage of both algorithms.

libpmemobj Unified Logging

Both redo and undo logging in `libpmemobj` share the same internal interface and data structure, which is called a unified log (or ulog for short). This is because redo and undo logging only differ in the execution order of the log phases, or more precisely, when the log is applied on commit or recovery. In practice, however, there are performance considerations that require specialization in certain parts of the algorithm.

The ulog data structure contains one cache line header with metadata and a variable length array of data bytes. The header consists of:

- A checksum for both the header and data, used only for redo logs

- A monotonically increasing generation number of a transaction in the log, used only for undo logs

- The total length in bytes of the data array

- An offset of the next log in the group

The last field is used to create a singly linked list of all logs that participate in a single transaction. This is because it is impossible to predict the total required size of the log at the beginning of the transaction, so the library cannot allocate a log structure that is the exact required length ahead of time. Instead, the logs are allocated on demand and atomically linked into a list.

The unified log supports two ways of fail-safe inserting of entries:

1. **Bulk insert** takes an array of log entries, prepares the header of the log, and creates a checksum of both the header and data. Once done, a non-temporal copy, followed by a fence, is performed to store this structure into persistent memory. This is the way in which a group of deferred memory modifications forms a redo log with only one additional barrier at the end of the transaction. In this case, the checksum in the header is used to verify the consistency of the entire log. If that checksum doesn't match, the log is skipped during recovery.

2. **Buffer insert** takes only a single entry, checksums it together with the current generation number, and stores it in persistent memory through non-temporal stores followed by a fence. This method is used to create undo logs when snapshotting. Undo logs

in a transaction are different than redo logs because during the commit's fast path, they need to be invalidated instead of applied. Instead of laboriously writing zeros into the log buffer, the log is invalidated by incrementing the generation number. This works because the number is part of the data with its checksum, so changing the generation number will cause a checksum failure. This algorithm allows libpmemobj to have only one additional fence for the transaction (on top of the fences needed for snapshots) to ensure fail-safety of a log, resulting in very low-overhead transactions.

Persistent Allocations: The Interface of a Transactional Persistent Allocator

The internal allocator interface in libpmemobj is far more complex than a typical volatile dynamic memory allocator. First, it must ensure fail-safety of all its operations and cannot allow for any memory to become unreachable due to interruptions. Second, it must be transactional so that multiple operations on the heap can be done atomically alongside other modifications. And lastly, it must operate on the pool state, allocating memory from specific files instead of relying on the anonymous virtual memory provided by the operating system. All these factors contribute to an internal API that hardly resembles the standard malloc() and free(), shown in Listing 16-1.

Listing 16-1. The core persistent memory allocator interface that splits heap operations into two distinct steps

```
int palloc_reserve(struct palloc_heap *heap, size_t size,...,
        struct pobj_action *act);
void palloc_publish(struct palloc_heap *heap,
        struct pobj_action *actv, size_t actvcnt,
        struct operation_context *ctx);
```

All memory operations, called "actions" in the API, are broken up into two individual steps.

The first step reserves the state that is needed to perform the operation. For allocations, this means retrieving a free memory block, marking it as reserved, and initializing the object's content. This reservation is stored in a user-provided runtime variable. The library guarantees that if an application crashes while holding reservations, the persistent state is not affected. That is why these action variables must not be persistent.

The second step is the act of exercising the reservations, which is called "publication." Reservations can be published individually, but the true power of this API lies in its ability to group and publish many different actions together.

The internal allocator API also has a function to create an action that will set a memory location to a given value when published. This is used to modify the destination pointer value and is instrumental in making the atomic API of `libpmemobj` fail-safe.

All internal allocator APIs that need to perform fail-safe atomic actions take operation context as an argument, which is the runtime instance of a single log. It contains various state information, such as the total capacity of the log and the current number of entries. It exposes the functions to create either bulk or singular log entries. The allocator's functions will log and then process all metadata modifications inside of the persistent log that belongs to the provided instance of the operating context.

Persistent Memory Heap Management: Allocator Design for Persistent Memory

The previous section described the interface for the memory allocation used internally in `libpmemobj`, but that was only the tip of the allocator iceberg. Before diving deeper into this topic, we briefly describe the principles behind normal volatile allocators so you can understand how persistent memory impacts the status quo.

Traditional allocators for volatile memory are responsible for efficient – in both time and space – management of operating system–provided memory pages. Precisely how this should be done for the generic case is an active research area of computer science; many different techniques can be used. All of them try to exploit the regularities in allocation and deallocation patterns to minimize heap fragmentation.

Most commonly used general-purpose memory allocators settled on an algorithm that we refer to as "segregated fit with page reuse and thread caching."

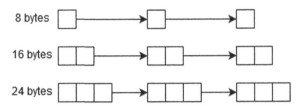

Figure 16-6. *Example of free lists in a memory allocator*

This works by using a free list for many different sizes, shown in Figure 16-6, until some predefined threshold, after which it is sensible to allocate directly from the operating system. Those free lists are typically called bins or buckets and can be implemented in various ways, such as a simple linked list or contiguous buffer with boundary tags. Each incoming memory allocation request is rounded up to match one of the free lists, so there must be enough of them to minimize the amount of overprovisioned space for each allocation. This algorithm approximates a best-fit allocation policy that selects the memory block with the least amount of excess space for the request from the ones available.

Using this technique allows memory allocators to have average-case O(1) complexity while retaining the memory efficiency of best fit. Another benefit is that rounding up of memory blocks and subsequent segregation forces some regularity to allocation patterns that otherwise might not exhibit any.

Some allocators also sort the available memory blocks by address and, if possible, allocate the one that is spatially collocated with previously selected blocks. This improves space efficiency by increasing the likelihood of reusing the same physical memory page. It also preserves temporal locality of allocated memory objects, which can minimize cache and translation lookaside buffer (TLB) misses.

One important advancement in memory allocators is scalability in multithreaded applications. Most modern memory allocators implement some form of thread caching, where the vast majority of allocation requests are satisfied directly from memory that is exclusively assigned to a given thread. Only when memory assigned to a thread is entirely exhausted, or if the request is very large, the allocation will contend with other threads for operating system resources.

This allows for allocator implementations that have no locks of any kind, not even atomics, on the fast path. This can have a potentially significant impact on performance, even in the single-threaded case. This technique also prevents allocator-induced false sharing between threads, since a thread will always allocate from its own region of memory. Additionally, the deallocation path often returns the memory block to the thread cache from which it originated, again preserving locality.

We mentioned earlier that volatile allocators manage operating system–provided pages but did not explain how they acquire those pages. This will become very important later as we discuss how things change for persistent memory. Memory is usually requested on demand from the operating system either through sbrk(), which moves the break segment of the application, or anonymous mmap(), which creates new virtual memory mapping backed by the page cache. The actual physical memory is usually not assigned until the page is written to for the first time. When the allocator decides that it no longer needs a page, it can either completely remove the mapping using unmap() or it can tell the operating system to release the backing physical pages but keep the virtual mapping. This enables the allocator to reuse the same addresses later without having to memory map them again.

How does all of this translate into persistent memory allocators and libpmemobj specifically?

The persistent heap must be resumable after application restart. This means that all state information must be either located on persistent memory or reconstructed on startup. If there are any active bookkeeping processes, those need to be restarted from the point at which they were interrupted. There cannot be any volatile state held in persistent memory, such as thread cache pointers. In fact, the allocator must not operate on any pointers at all because the virtual address of the heap can change between restarts.

In libpmemobj, the heap is rebuilt lazily and in stages. The entire available memory is divided into equally sized zones (except for the last one, which can be smaller than the others) with metadata at the beginning of each one. Each zone is subsequentially divided into variably sized memory blocks called chunks. Whenever there is an allocation request, and the runtime state indicates that there is no memory to satisfy it, the zone's metadata is processed, and the corresponding runtime state is initialized. This minimizes the startup time of the pool and amortizes the cost of rebuilding the heap state across many individual allocation requests.

There are three main reasons for having any runtime state at all. First, access latency of persistent memory can be higher than that of DRAM, potentially impacting performance of data structures placed on it. Second, separating the runtime state from the persistent state enables a workflow where the memory is first reserved in runtime state and initialized, and only then the allocation is reflected on the persistent state.

This mechanism was described in the previous section. Finally, maintaining fail-safety of complex persistent data structures is expensive, and keeping them in DRAM allows the allocator to sidestep that cost.

The runtime allocation scheme employed by libpmemobj is segregated fit with chunk reuse and thread caching as described earlier. Free lists in libpmemobj, called buckets, are placed in DRAM and are implemented as vectors of pointers to persistent memory blocks. The persistent representation of this data structure is a bitmap, located at the beginning of a larger buffer from which the smaller blocks are carved out. These buffers in libpmemobj, called runs, are variably sized and are allocated from the previously mentioned chunks. Very large allocations are directly allocated as chunks. Figure 16-7 shows the libpmemobj implementation.

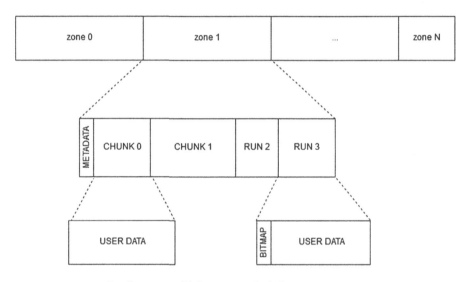

Figure 16-7. *On-media layout of libpmemobj's heap*

Persistent allocators must also ensure consistency in the presence of failures, otherwise, memory might become unreachable after an ungraceful shutdown of the application. One part of the solution is the API we outlined in the previous section. The other part is the careful design of the algorithms inside the allocator that ensures no matter when the application is aborted, the state is consistent. This is also aided by redo logs, which are used to ensure atomicity of groups of noncontiguous persistent metadata changes.

One of the most impactful aspects of persistent memory allocation is how the memory is provisioned from the operating system. We previously explained that for normal volatile allocators, the memory is usually acquired through anonymous memory mappings that are backed by the page cache. In contrast, persistent heaps must use file-based memory mappings, backed directly by persistent memory. The difference might be subtle, but it has a significant impact on the way the allocator must be designed. The allocator must manage the entire virtual address space, retain information about any potential noncontiguous regions of the heap, and avoid excessive overprovisioning of virtual address space. Volatile allocators can rely on the operating system to coalesce noncontiguous physical pages into contiguous virtual ones, whereas persistent allocators cannot do the same without explicit and complicated techniques. Additionally, for some file system implementations, the allocator cannot assume that the physical memory is allocated at the time of the first page fault, so it must be conservative with internal block allocations.

Another problem for allocation from file-based mappings is that of perception. Normal allocators, due to memory overcommitment, seemingly never run out of memory because they are allocating the virtual address space, which is effectively infinite. There are negative performance consequences of address space bloat, and memory allocators actively try to avoid it, but they are not easily measurable in a typical application. In contrast, memory heaps allocate from a finite resource, the persistent memory device, or a file. This exacerbates the common phenomenon that is heap fragmentation by making it trivially measurable, creating the perception that persistent memory allocators are less efficient than volatile ones. They can be, but the operating system does a lot of work behind the scene to hide fragmentation of traditional memory allocators.

ACID Transactions: Efficient Low-Level Persistent Transactions

The four components we just described – lanes, redo logs, undo logs, and the transactional memory allocator – form the basis of libpmemobj's implementation of ACID transactions that we defined in Chapter 4.

A transaction's persistent state consists of three logs. First is an undo log, which contains snapshots of user data. Second is an external redo log, which contains allocations and deallocations performed by the user. Third is an internal redo log, which is used to perform atomic metadata allocations and deallocations. This is technically not part of the transaction but is required to allocate the log extensions if they are needed. Without the internal redo log, it would be impossible to reserve and then publish a new log object in a transaction that already had user-made allocator actions in the external redo log.

All three logs have individual operation-context instances that are stored in runtime state of the lanes. This state is initialized when the pool is opened, and that is also when all the logs of the prior instance of the application are either processed or discarded. There is no special persistent variable that indicates whether past transactions in the log were successful or not. That information is directly derived from checksums stored in the log.

When a transaction begins, and it is not a nested transaction, it acquires a lane, which must not contain any valid uncommitted logs. The runtime state of the transaction is stored in a thread-local variable, and that is where the lane variable is stored once acquired.

Transactional allocator operations use the external redo log and its associated operation context to call the appropriate reservation method which in turn creates an allocator action to be published at the time of transaction commit. The allocator actions are stored in a volatile array. If the transaction is aborted, all the actions are canceled, and the associated state is discarded. The complete redo log for memory allocations is created only at the time of transaction commit. If the library is interrupted while creating the redo log, the next time the pool is opened, the checksum will not match, and the transaction will be aborted by rolling back using the undo log.

Transactional snapshots use the undo log and its context. The first time a snapshot is created, a new memory modification action is created in the external redo log. When published, that action increments the generation number of the associated undo log, invalidating its contents. This guarantees that if the external log is fully written and processed, it automatically discards the undo log, committing the entire transaction. If the external log is discarded, the undo log is processed, and the transaction is aborted.

To ensure that there are never two snapshots of the same memory location (this would be an inefficient use of space), there is a runtime range tree that is queried every time the application wants to create an undo log entry. If the new range overlaps with an existing snapshot, adjustments to the input arguments are made to avoid duplication. The same mechanism is also used to prevent snapshots of newly allocated data. Whenever new memory in a transaction is allocated, the reserved memory range is inserted into the ranges tree. Snapshotting new objects is redundant because they will be discarded automatically in the case of an abort.

To ensure that all memory modifications performed inside the transaction are durable on persistent memory once committed, the ranges tree is also used to iterate over all snapshots and call the appropriate flushing function on the modified memory locations.

Lazy Reinitialization of Variables: Storing the Volatile State on Persistent Memory

While developing software for persistent memory, it is often useful to store the runtime (volatile) state inside of persistent memory locations. Keeping that state consistent, however, is extremely difficult, especially in multithreaded applications.

The problem is the initialization of the runtime state. One solution is to simply iterate over all objects at the start of the application and initialize the volatile variables then, but that might significantly contribute to startup time of applications with large persistent pools. The other solution is to lazily reinitialize the variables on access, which is what libpmemobj does for its built-in locks. The library also exposes this mechanism through an API for use with custom algorithms.

Lazy reinitialization of the volatile state is implemented using a lock-free algorithm that relies on a generation number stored alongside each volatile variable on persistent memory and inside the pool header. The pool header resident copy is increased by two every time a pool is opened. This means that a valid generation number is always even. When a volatile variable is accessed, its generation number is checked against the one stored in the pool header. If they match, it means that the object can be used and is simply returned to the application; otherwise, the object needs to be initialized before returning to ensure the initialization is thread-safe and is performed exactly once in a single instance of the application.

The naive implementation could use a double-checked locking, where a thread would try to acquire a lock prior to initialization and verify again if the generation numbers match. If they still do not match, initialize the object, and increase the number. To avoid the overhead that comes with using locks, the actual implementation first uses a compare-and-swap to set the generation number to a value that is equal to the generation number of the pool minus one, which is an odd number that indicates an initialization operation is in progress. If this compare-and-swap were to fail, the algorithm would loop back to check if the generation number matches. If it is successful, the running thread initializes the variable and once again increments the generation number – this time to an even number that should match the number stored in the pool header.

Summary

This chapter described the architecture and inner workings of `libpmemobj`. We also discuss the reasons for the choices that were made during the design and implementation of `libpmemobj`. With this knowledge, you can accurately reason about the semantics and performance characteristics of code written using this library.

## CHAPTER 17

Reliability, Availability, and Serviceability (RAS)

This chapter describes the high-level architecture of reliability, availability, and serviceability (RAS) features designed for persistent memory. Persistent memory RAS features were designed to support the unique error-handling strategy required for an application when persistent memory is used. Error handling is an important part of the program's overall reliability, which directly affects the availability of applications. The error-handling strategy for applications impacts what percentage of the expected time the application is available to do its job.

Persistent memory vendors and platform vendors will both decide which RAS features and how they will be implemented at the lowest hardware levels. Some common RAS features were designed and documented in the ACPI specification, which is maintained and owned by the UEFI Forum (`https://uefi.org/`). In this chapter, we try to attain a general perspective of these ACPI-defined RAS features and call out vendor-specific details if warranted.

Dealing with Uncorrectable Errors

The main memory of a server is protected using error correcting codes (ECC). This is a common hardware feature that can automatically correct many memory errors that happen due to transient hardware issues, such as power spikes, soft media errors, and so on. If an error is severe enough, it will corrupt enough bits that ECC cannot correct; the result is called an uncorrectable error (UE).

Uncorrectable errors in persistent memory require special RAS handling that differs from how a platform may traditionally handle volatile memory uncorrectable errors.

© The Author(s) 2020
S. Scargall, *Programming Persistent Memory*, https://doi.org/10.1007/978-1-4842-4932-1_17

Persistent memory uncorrectable errors are *persistent*. Unlike volatile memory, if power is lost or an application crashes and restarts, the uncorrectable error will remain on the hardware. This can lead to an application getting stuck in an infinite loop such as

1. Application starts

2. Reads a memory address

3. Encounters uncorrectable error

4. Crashes (or system crashes and reboots)

5. Starts and resumes operation from where it left off

6. Performs a read on the same memory address that triggered the previous restart

7. Crashes (or system crashes and reboots)

8. ...

9. Repeats infinitely until manual intervention

The operating system and applications may need to address uncorrectable errors in three main ways:

- When consuming previously undetected uncorrectable errors during runtime

- When unconsumed uncorrectable errors are detected at runtime

- When mitigating uncorrectable memory locations detected at boot

Consumed Uncorrectable Error Handling

When an uncorrectable error is detected on a requested memory address, data poisoning is used to inform the CPU that the data requested has an uncorrectable error. When the hardware detects an uncorrectable memory error, it routes a poison bit along with the data to the CPU. For the Intel architecture, when the CPU detects this poison bit, it sends a processor interrupt signal to the operating system to notify it of this error. This signal is called a machine check exception (MCE). The operating system can then examine the uncorrectable memory error, determine if the software can recover, and perform recovery actions via an MCE handler. Typically, uncorrectable errors fall into three categories:

- Uncorrectable errors that may have corrupted the state of the CPU and require a system reset.

- Uncorrectable errors that can be recovered by software can be handled during runtime.

- Uncorrectable errors that require no action.

Operating system vendors handle this uncorrectable error notification in different ways, but some common elements exist for all of them.

Using Linux as an example, when the operating system receives a processor interrupt for an uncorrectable error, it proceeds to offline the page of memory where the uncorrectable error occurred and add the error to a list of areas containing known uncorrectable errors. This list of known uncorrectable errors is called the bad block list. Linux will also mark the page that contains the uncorrectable error to be cleared when the page is recycled for use by another application.

The PMDK libraries automatically check the list of pages with uncorrectable errors in the operating system and prevent an application from opening a persistent memory pool if it contains errors. If a page of memory is in use by an application, Linux attempts to kill it using the SIGBUS mechanism.

At this point, the application developer can decide what to do with this error notification. The simplest way for you to handle uncorrectable errors is to let the application die when it gets a SIGBUS so you do not need to write the complicated logic of handling a SIGBUS at runtime. Instead, on restart, the application can use PMDK to detect that the persistent memory pool contains errors and repair the data during application initialization. For many applications, this repair can be as simple as reverting to a backup error-free copy of the data.

Figure 17-1 shows a simplified sequence of how Linux can handle an uncorrectable (but not fatal) error that was consumed by an application.

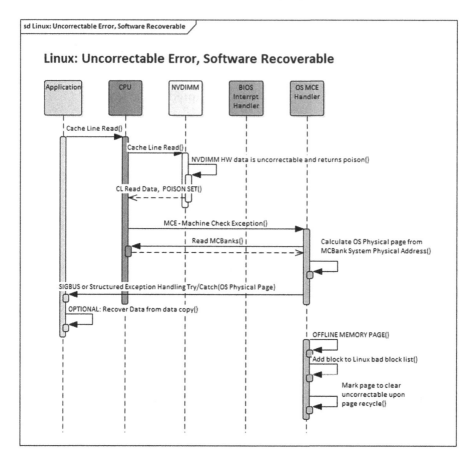

Figure 17-1. *Linux consumed uncorrectable error-handling sequence*

Unconsumed Uncorrectable Error Handling

RAS features are defined to inform software of uncorrectable errors that have been discovered on the persistent memory media but have not yet been consumed by software. The goal of this feature is to allow the operating system to opportunistically offline or clear pages with known uncorrectable errors before they can be used by an application. If the address of the uncorrectable error is already in use by an application, the operating system may also choose to notify it of the unconsumed uncorrectable error or wait until the application consumes the error. The operating system may choose to wait on the chance that the application never tries to access the affected page and later return the page to the operating system for recycling. At this time, the operating system would clear or offline the uncorrectable error.

Unconsumed uncorrectable error handling may be implemented differently on different vendor platforms, but at the core, there will always be a mechanism to discover the unconsumed uncorrectable error, a mechanism to signal the operating system of an unconsumed uncorrectable error, and a mechanism for the operating system to query information about the unconsumed uncorrectable error. As shown in Figure 17-2, these three mechanisms work together to proactively keep the operating system informed of all discovered uncorrectable errors during runtime.

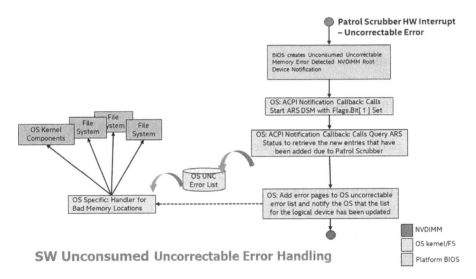

Figure 17-2. *Unconsumed uncorrectable error handling*

Patrol Scrub

Patrol scrub (also known as memory scrubbing) is a long-standing RAS feature for volatile memory that can also be extended to persistent memory. It is an excellent example of how a platform can discover uncorrectable errors in the background during normal operation.

Patrol scrubbing is done using a hardware engine, on either the platform or on the memory device, which generates requests to memory addresses on the memory device. The engine generates memory requests at a predefined frequency. Given enough time, it will eventually access every memory address. The frequency in which patrol scrub generates requests produces no noticeable impact on the memory device's quality of service.

By generating read requests to memory addresses, the patrol scrubber allows the hardware an opportunity to run ECC on a memory address and correct any correctable errors before they can become uncorrectable errors. Optionally, if an uncorrectable error is discovered, the patrol scrubber can trigger a hardware interrupt and notify the software layer of its memory address.

Unconsumed Uncorrectable Memory-Error Persistent Memory Root-Device Notification

The ACPI specification describes a method for hardware to notify software of unconsumed uncorrectable errors called the Unconsumed Uncorrectable Memory-Error Persistent Memory Root-Device Notification. Using the ACPI-defined framework, the operating system can subscribe to be notified by the platform whenever an uncorrectable memory error is detected. It is the platform's responsibility to receive notification from persistent memory devices that an uncorrectable error has been detected and take appropriate action to generate a persistent memory root-device notification. Upon receipt of root-device notification, the operating system can then use existing ACPI methods, such as Address Range Scrub (ARS), to discover the address of the newly created uncorrectable memory error and take appropriate actions.

Address Range Scrub

ARS is a device-specific method (_DSM) defined in the ACPI specification. Privileged software can call an ACPI _DSM such as ARS at runtime to retrieve or scan for the locations of uncorrectable memory errors for all persistent memory in the platform. Because ARS is implemented by the platform, each vendor may implement some of the functionality differently.

An ARS accepts a given system address range from the caller and, like patrol scrub, inspects each memory address in that range for memory errors. When ARS completes, the caller is given a list of memory addresses in the given range that contains memory errors. Inspection of each memory address may be handled by persistent memory hardware or by the platform itself. Unlike a patrol scrub, ARS inspects each memory address at a very high frequency. This increased frequency of the scrub may impact the quality of service for the persistent memory hardware. Thus, ARS can optionally be invoked by the caller to return the results of the previous ARS, sometimes referred to as a short ARS.

Traditionally, the operating system executes ARS in one of two ways to obtain the addresses of uncorrectable errors after a boot. Either a full scan is executed on all the available persistent memory during system boot or after an unconsumed uncorrectable memory error root-device notification is received. In both instances, the intent is to discover these addresses before they are consumed by applications.

Operating systems will compare the list of uncorrectable errors returned by ARS to their persistent list of uncorrectable errors. If new errors are detected, the list is updated. This list is intended to be consumed by higher-level software, such as the PMDK libraries.

Clearing Uncorrectable Errors

Uncorrectable errors for persistent memory will survive power loss and may require special handling to clear corrupted data from the memory address. When an uncorrectable error is cleared, the data at the requested memory address is modified, and the error is cleared. Because hardware cannot silently modify application data, clearing uncorrectable errors is the software's responsibility. Clearing uncorrectable errors is optional, and some operating systems may choose to only offline memory pages that contain memory errors instead of recycling memory pages that contain uncorrectable errors. In some operating systems, privileged applications may have access to clear uncorrectable errors. Nevertheless, an operating system is not required to provide this access.

The ACPI specification defines a Clear Uncorrectable Error DSM for operating systems to instruct the platform to clear the uncorrectable errors. While persistent memory programming is byte addressable, clearing uncorrectable errors is not. Different vendor implementations of persistent memory may specify the alignment and size of the memory unit that is to be cleared by a Clear Uncorrectable Error. Any internal platform or operating system list of memory errors should also be updated upon successful executing of the Clear Uncorrectable Error DSM command.

Device Health

System administrators may wish to act and mitigate any device health issues before they begin to affect the availability of applications using persistent memory. To that end, operating systems or management applications will want to discover an accurate picture of persistent memory device health to correctly determine the reliability of the persistent memory. The ACPI specification defines a few vendor-agnostic health discovery methods, but many vendors choose to implement additional persistent memory device methods for attributes that are not covered by the vendor-agnostic methods. Many of these vendor-specific health discovery methods are implemented as an ACPI device-specific method (_DSM). Applications should be aware of degradation to the quality of service if they call ACPI methods directly, since some platform implementations may impact memory traffic when ACPI methods are invoked. Avoid excessive polling of device health methods when possible.

On Linux, the ndctl utility can be used to query the device health of persistent memory modules. Listing 17-1 shows an example output of an Intel Optane DC persistent memory module.

Listing 17-1. Using ndctl to query the health of persistent memory modules

```
$ sudo ndctl list -DH -d nmem1
[
  {
    "dev":"nmem1",
    "id":"8089-a2-1837-00000bb3",
    "handle":17,
    "phys_id":44,
    "security":"disabled",
    "health":{
      "health_state":"ok",
      "temperature_celsius":30.0,
      "controller_temperature_celsius":30.0,
      "spares_percentage":100,
      "alarm_temperature":false,
      "alarm_controller_temperature":false,
      "alarm_spares":false,
      "alarm_enabled_media_temperature":false,
      "alarm_enabled_ctrl_temperature":false,
      "alarm_enabled_spares":false,
      "shutdown_state":"clean",
      "shutdown_count":1
    }
  }
]
```

Conveniently, ndctl also provides a monitoring command and daemon to continually monitor the health of the systems' persistent memory modules. For a list of all the available options, refer to the ndctl-monitor(1) man page. Examples for using this monitor method include

Example 1: Run a monitor as a daemon to monitor DIMMs on bus "nfit_test.1,"

```
$ sudo ndctl monitor --bus=nfit_test.1 --daemon
```

Example 2: Run a monitor as a one-shot command, and output the notifications to /var/log/ndctl.log.

```
$ sudo ndctl monitor --log=/var/log/ndctl.log
```

Example 3: Run a monitor daemon as a system service.

```
$ sudo systemctl start ndctl-monitor.service
```

You can obtain similar information using the persistent memory device-specific utility. For example, you can use the ipmctl utility on Linux and Windows∗ to obtain hardware-level data similar to that shown by ndctl. Listing 17-2 shows health information for DIMMID 0x0001 (nmem1 equivalent in ndctl terms).

Listing 17-2. Health information for DIMMID 0x0001

```
$ sudo ipmctl show -sensor -dimm 0x0001

DimmID | Type                        | CurrentValue
========================================================
0x0001 | Health                      | Healthy
0x0001 | MediaTemperature            | 30C
0x0001 | ControllerTemperature       | 31C
0x0001 | PercentageRemaining         | 100%
0x0001 | LatchedDirtyShutdownCount   | 1
0x0001 | PowerOnTime                 | 27311231s
0x0001 | UpTime                      | 6231933s
0x0001 | PowerCycles                 | 170
0x0001 | FwErrorCount                | 8
0x0001 | UnlatchedDirtyShutdownCount | 107
```

ACPI-Defined Health Functions (_NCH, _NBS)

The ACPI specification includes two vendor-agnostic methods for operating systems and management software to call for determining the health of a persistent memory device.

Get NVDIMM Current Health Information (_NCH) can be called by the operating systems at boot time to get the current health of the persistent memory device and take appropriate action. The values reported by _NCH can change during runtime and should be monitored for changes. _NCH contains health information that shows if

- The persistent memory requires maintenance

- The persistent memory device performance is degraded

- The operating system can assume write persistency loss on subsequent power events

- The operating system can assume all data will be lost on subsequent power events

Get NVDIMM Boot Status (_NBS) allows operating systems a vendor-agnostic method to discover persistent memory health status that does not change during runtime. The most significant attribute reported by _NBS is Data Loss Count (DLC). Data Loss Count is expected to be used by applications and operating systems to help identify the rare case where a persistent memory dirty shutdown has occurred. See "Unsafe/Dirty Shutdown" later in this chapter for more information on how to properly use this attribute.

Vendor-Specific Device Health (_DSMs)

Many vendors may want to add further health attributes beyond what exists in _NBS and _NCH. Vendors are free to design their own ACPI persistent memory device-specific methods (_DSM) to be called by the operating system and privileged applications. Although vendors implement persistent memory health discovery differently, a few common health attributes are likely to exist to determine if a persistent memory device requires service. These health attributes may include information such as an overall health summary of the persistent memory, current persistent memory temperature, persistent media error counts, and total device lifetime utilization. Many operating systems, such as Linux, include support to retrieve and report the vendor-unique health statistics through tools such as ndctl. The Intel persistent memory _DSM interface document can be found under the "Related Specification" section of https://docs.pmem.io/.

ACPI NFIT Health Event Notification

Due to the potential loss of quality of service, operating systems and privileged applications may not want to actively poll persistent memory devices to retrieve device health. Thus, the ACPI specification has defined a passive notification method to allow the persistent memory device to notify when a significant change in device health has occurred. Persistent memory device vendors and platform BIOS vendors decide which device health changes are significant enough to trigger an NVDIMM Firmware Interface Table (NFIT) health event notification. Upon receipt of an NFIT health event, a notification to the operating system is expected to call an _NCH or a _DSM attached to the persistent memory device and take appropriate action based on the data returned.

Unsafe/Dirty Shutdown

An unsafe or dirty shutdown on persistent memory means that the persistent memory device power-down sequence or platform power-down sequence may have failed to write all in-flight data from the system's persistence domain to persistent media. (Chapter 2 describes persistence domains.) A dirty shutdown is expected to be a very rare event, but they can happen due to a variety of reasons such as physical hardware issues, power spikes, thermal events, and so on.

A persistent memory device does not know if any application data was lost as a result of the incomplete power-down sequence. It can only detect if a series of events occurred in which data may have been lost. In the best-case scenario, there might not have been any applications that were in the process of writing data when the dirty shutdown occurred.

The RAS mechanism described here requires the platform BIOS and persistent memory vendor to maintain a persistent rolling counter that is incremented anytime a dirty shutdown is detected. The ACPI specification refers to such a mechanism as the Data Loss Count (DLC) that can be returned as part of the Get NVDIMM Boot Status(_NBS) persistent memory device method.

Referring to the output from `ndctl` in Listing 17-1, the `"shutdown_count"` is reported in the health information. Similarly, the output from `ipmctl` in Listing 17-2 reports `"LatchedDirtyShutdownCount"` as the dirty shutdown counter. For both outputs, a value of 1 means no issues were detected.

Application Utilization of Data Loss Count (DLC)

Applications may want to use the DLC counter provided by _NBS to detect if possible data loss occurred while saving data from the system's persistence domain to the persistent media. If such a loss can be detected, applications can perform data recovery or rollback using application-specific features.

The application's responsibilities and possible implementation suggestions for applications are outlined as follows:

1. Application first creates its initial metadata and stores it in a persistent memory file:

   a. Application retrieves current DLC via operating system–specific means for the physical persistent memory that make up the logical volume the applications metadata resides on.

   b. Application calculates the current Logical Data Loss Count (LDLC) as the sum of the DLC for all physical persistent memory that make up the logical volume the applications metadata resides on.

   c. Application stores the current LDLC in its metadata file and ensures that the update of the LDLC has been flushed to the system's persistence domain. This is done by using a flush that forces the write data all the way to the persistent memory power-fail safe domain. (Chapter 2 contains more information about flushing data to the persistence domain.)

   d. Application determines GUID or UUID for the logical volume the applications metadata resides on, stores this in its metadata file, and ensures the update of the GUID/UUID to the persistence domain. This is used by the application to later identify if the metadata file has been moved to another logical volume, where the current DLC is no longer valid.

   e. Application creates and sets a "clean" flag in its metadata file and ensures the update of the clean flag to the persistence domain. This is used by the application to determine if the application was actively writing data to persistence during dirty shutdown.

2. Every time the application runs and retrieves its metadata from persistent memory:

   a. Application checks the GUID/UUID saved in its metadata with the current UUID for the logical volume the applications metadata resides on. If they match, then the LDLC is describing the same logical volume the app was using. If they do not match, then the DLC is for some other logical volume and no longer applies. The application decides how to handle this.

   b. Application calculates the current LDLC as the sum of the DLC for all physical persistent memory the application's metadata resides on.

   c. Application compares the current LDLC calculated with the saved LDLC retrieved from its metadata.

   d. If the current LDLC does not match the saved LDLC, then one or more persistent memory have detected a dirty shutdown and possible data loss. If they do match, no further action is required by the application.

   e. Application checks the status of the saved "clean" flag in its metadata; if the clean flag is NOT set, this application was writing at the time of the shutdown failure.

   f. If the clean flag is NOT set, perform software data recovery or rollback using application-specific functionality.

   g. Application stores the new current LDLC in its metadata file and ensures that the update of the count has been flushed to the system's persistence domain. This may require unsetting the clean flag if it was previously set.

   h. Application sets the clean flag in its metadata file and ensures that the update of the clean flag has been flushed to the persistence domain.

3.   Every time the application will write to the file:

a. Before the application writes data, it clears the "clean" flag in its metadata file and ensures that the flag has been flushed to the persistence domain.

b. Application writes data to its persistent memory space.

c. After the application completes writing data, it sets the "clean" flag in its metadata file and ensures the flag has been flushed to the persistence domain.

PMDK libraries make these steps significantly easier and account for interleaving set configurations.

Summary

This chapter describes some of the RAS features that are available to persistent memory devices and that are relevant to persistent memory applications. It should have given you a deeper understanding of uncorrectable errors and how applications can respond to them, how operating systems can detect health status changes to improve the availability of applications, and how applications can best detect dirty shutdowns and use the data loss counter.

CHAPTER 18

Remote Persistent Memory

This chapter provides an overview of how persistent memory – and the programming concepts that were introduced in this book – can be used to access persistent memory located in remote servers connected via a network. A combination of TCP/IP or RDMA network hardware and software running on the servers containing persistent memory provide direct remote access to persistent memory.

Having remote direct memory access via a high-performance network connection is a critical use case for most cloud deployments of persistent memory. Typically, in high-availability or highly redundant use cases, data written locally to persistent memory is not considered reliable until it has been replicated to two or more remote persistent memory devices on separate remote servers. We describe this push model design later in this chapter.

While it is certainly possible to use existing TCP/IP networking infrastructures to remotely access the persistent memory, this chapter focuses on the use of remote direct memory access (RDMA). Direct memory access (DMA) allows data movement on a platform to be off-loaded to a hardware DMA engine that moves that data on behalf of the CPU, freeing it to do other important tasks during the data move. RDMA applies the same concept and enables data movement between remote servers to occur without the CPU on either server having to be directly involved.

This chapter's content and the PMDK `librpmem` remote persistent memory library that is discussed assume the use of RDMA, but the concepts discussed here can apply to other networking interconnects and protocols.

© The Author(s) 2020
S. Scargall, *Programming Persistent Memory*, https://doi.org/10.1007/978-1-4842-4932-1_18

Figure 18-1 outlines a simple remote persistent memory configuration with one initiator system that is replicating writes to persistent memory on a single remote target system. While this shows the use of persistent memory on both the initiator and target, it is possible to read data from initiator DRAM and write to persistent memory on the remote target system, or read from the initiator's persistent memory and write to the remote target's DRAM.

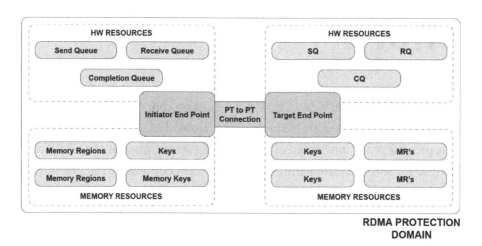

Figure 18-1. *Initiator and target system using RDMA*

RDMA Networking Protocols

Examples of popular RDMA networking protocols used throughout the cloud and enterprise data centers include:

- InfiniBand is an I/O architecture and high-performance specification for data transmission between high-speed, low-latency, and highly scalable CPUs, processors, and storage.

- RoCE (RDMA over Converged Ethernet) is a network protocol that allows RDMA over an Ethernet network.

- iWARP (Internet Wide Area RDMA Protocol) is a networking protocol that implements RDMA for efficient data transfer over Internet Protocol networks.

All three protocols support high-performance data movement to and from persistent memory using RDMA.

The RDMA protocols are governed by the RDMA Wire Protocol Standards, which are driven by the IBTA (InfiniBand Trade Association) and the IEFT (Internet Engineering Task Force) specifications. The IBTA (`https://www.infinibandta.org/`) governs the InfiniBand and RoCE protocols, while the IEFT (`https://www.ietf.org/`) governs iWARP.

Low-latency RDMA networking protocols allow the network interface controller (NIC) to control the movement of data between an initiator node source buffer and the sink buffer on the target node without needing either node's CPU to be involved in the data movement. In fact, RDMA Read and RDMA Write operations are often referred to as one-sided operations because all of the information required to move the data is supplied by the initiator and the CPU on the target node is not typically interrupted or even aware of the data transfer.

To perform remote data transfers, information from the target node's buffers must be passed to the initiator before the remote operation(s) will begin. This requires configuring the local initiator's RDMA resources and buffers. Similarly, the remote target node's RDMA resources that will require CPU resources will need to be initialized and reported to the initiator. However, once the resources for the RDMA transfers are set up and applications initiate the RDMA request using the CPU, the NIC does the actual data movement on behalf of the RDMA-aware application.

RDMA-aware applications are responsible for:

- Interrogating each NIC on every initiator and target system to determine supported features

- Selecting a NIC for each end of the RDMA point-to-point connection

- Creating the connection with the selected NICs, described as an RDMA protection domain

- Allocating queues for the incoming and outgoing message on each NIC and assigning those hardware resources to the protection domain

- Allocating DRAM or persistent memory buffers for use with RDMA, registering those buffers with the NIC, and assigning those buffers to the protection domain

Three basic RDMA commands are used by most RDMA-capable applications and libraries:

RDMA Write: A one-sided operation where only the initiator supplies all of the information required for the transfer to occur. This transfer is used to write data to the remote target node. The write request contains all source and sink buffer information. The remote target system is not typically interrupted and thus completely unaware of the write operations occurring through the NIC. When the initiator's NIC sends a write to the target, it will generate a "software write completion interrupt." A *software write completion interrupt* means that the write message has been sent to the target NIC and is not an indicator of the write completion. Optionally, RDMA Writes can use an immediate option that will interrupt the target node CPU and allow software running there to be immediately notified of the write completion.

RDMA Read: A one-sided operation where only the initiator supplies all of the information required for the transfer to occur. This transfer is used to read data from the remote target node. The read request contains all source buffer and target sink buffer information, and the remote target system is not typically interrupted and thus completely unaware of the read operations occurring through the NIC. The initiator *software read completion interrupt* is an acknowledgment that the read has traversed all the way through the initiator's NIC, over the network, into the target system's NIC, through the target internal hardware mesh and memory controllers, to the DRAM or persistent memory to retrieve the data. Then it returns all the way back to the initiator software that registered for the completion notification.

RDMA Send (and Receive): The two-sided RDMA Send means that both the initiator and target must supply information for the transfer to complete. This is because the target NIC will be interrupted when the RDMA Send is received by the target NIC and requires a hardware receive queue to be set up and pre-populated with completion entries before the NIC will receive an RDMA Send transfer operation. Data from the initiator application is bundled in a small, limited sized buffer and sent to the target NIC. The target CPU will be interrupted to handle the send operation and any data it contains. If the initiator needs to be notified of receipt of the RDMA Send, or to handle a message back to the initiator, another RDMA Send operation must be sent in the reverse direction after the initiator has set up its own receive queue and queued completion entries to it. The use of the RDMA Send command and the contents of the payload are application-specific implementation details. An RDMA Send is typically used for bookkeeping and updates of read and write activity between the initiator and the target, since the target application has no other context of what data movement has taken place. For example, because there is no good way to know when writes have completed on the target, an RDMA Send is often used to notify the target node what is happening. For small amounts of data, the RDMA Send is very efficient, but it always requires target-side interaction to complete. An RDMA Write with immediate data operation will also allow the target node to be interrupted when the write has completed as a different mechanism for bookkeeping.

Goals of the Initial Remote Persistent Memory Architecture

The goal of the first remote persistent memory implementation was based on minimal changes – or ideally, no changes – to the current RDMA hardware and software stacks used with volatile memory. From a network hardware, middleware, and software architecture standpoint, writing to remote volatile memory is identical to writing to remote persistent memory. The knowledge that a specific memory-mapped file is backed by persistent memory vs. volatile memory is entirely the responsibility of the application to maintain. None of the lower layers in the networking stack are aware of the fact that the write is to a persistent memory region or volatile memory. The responsibility of knowing which write persistence method to use for a given target connection, and making those remote writes persistent, falls to the application.

Guaranteeing Remote Persistence

Until this chapter, much of the book focuses on the use and programming of persistent memory on the local machine. You are now aware of some of the challenges of using persistent memory, the persistence domain, and the need to understand and use a flushing mechanism to ensure the data is persistent. These same programming concepts and challenges apply to remote persistent memory with the additional constraints of making it work within the existing network protocol and network latency.

The SNIA NVM programming model (described in Chapter 3) requires applications to flush data that has been written to persistent memory to guarantee that the written data made it into the persistence domain. This same requirement applies to writing to remote persistent memory. After the RDMA Write or Send operation has moved the data from the initiator node to the persistent memory on the target node, that write or send data needs to be flushed to the persistence domain on the remote system. Alternatively, the remote write or send data needs to bypass CPU caches on the remote node to avoid having to be flushed.

Different vendor-specific platform features add an extra challenge to RDMA and to remote persistent memory. Intel platforms typically use a feature called allocating writes or Direct Data IO (DDIO) which allows incoming writes to be placed directly into the CPU's L3 cache. The data is immediately visible to any application wanting to read the data. However, having allocating writes enabled means that RDMA Writes to persistent memory now have to be flushed to the persistence domain on the target node.

On Intel platforms, allocating writes can be disabled by turning on non-allocating write I/O flows which forces the write data to bypass cache and be placed directly into the persistent memory, governed by the location of the RDMA Write sink buffer. This would slow down applications that will immediately touch the newly written data because they incur the penalty to pull the data into CPU cache. However, this simplifies making remote writes to persistent memory simpler and faster because cache flushing on the remote target node can be avoided. An additional complication to using non-allocating write mode on an Intel platform is that an entire PCI root complex must be enabled for this write mode. This means that any inbound writes that come through that PCI root complex, for any device connected downstream of it, will have write-data bypass CPU caches, causing possible additional performance latency as a side effect.

Intel specifies two methods for forcing writes to remote persistent memory into the persistence domain:

1. A general-purpose remote replication method that does not rely on Intel non-allocating write mode and assumes some or all of the remote write data will end up in CPU cache on the target system

2. A high-performance appliance remote replication method that uses the Intel platform-specific non-allocating write mode and is probably more suited to an appliance product where there is complete control over the hardware configuration to control what is connected to which PCI root complex

General-Purpose Remote Replication Method

The general-purpose remote replication method (GPRRM), also referred to as the general-purpose server persistency method (GPSPM), relies on the initiator RDMA application to maintain a list of virtual addresses on the remote target system that have been written to with previous RDMA Write requests. When all remote writes to persistent memory are issued, the application issues an RDMA Send request from the initiator NIC to the target NIC. The RDMA Send request contains a list of virtual starting addresses and lengths that the target system will consume when the application software running on the target node interrupts the system to process the send request. The application walks the list of regions, flushing each cache line in the requested region to the persistent memory using an optimized flush machine instruction (CLWB, CLFLUSHOPT, etc.). When complete, an SFENCE machine instruction is required to fence those previous writes and force them to complete before handling additional writes. The application on the target system then issues an RDMA Send request back to interrupt the initiator software of the completed flush operations. This is an indicator to the application that the previous writes were made persistent.

Figure 18-2 outlines the general-purpose remote replication method sequence of operation.

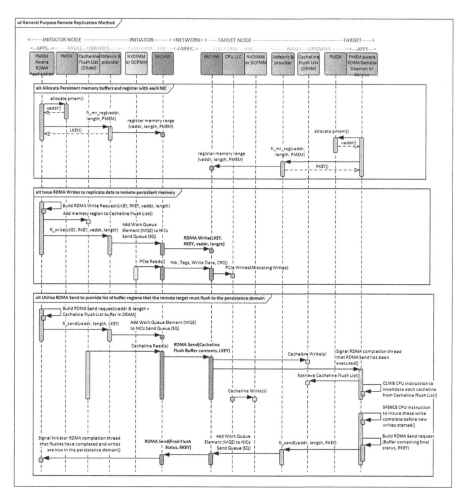

Figure 18-2. *The general-purpose remote replication method*

How Does the General-Purpose Remote Replication Method Make Data Persistent?

After the RDMA Write or any number of writes have been sent, the write data will either be in the L3 CPU cache (due to the default allocating writes) or persistent memory (assuming it does not all fit in L3) with potentially some write data still pending in NIC internal buffers. An RDMA Send request, by definition, will force previous writes to be pushed out of the NIC to the target L3 CPU cache and interrupt the target CPU. At this point, all previously issued RDMA Writes to persistent memory are now in L3 or persistent memory. The RDMA Send request contains a list of cache lines that the initiator is requesting the target system to flush to its persistence domain. The target system issues `optimized flush` instructions to flush each cache line in the list to the persistence domain. This is followed by an `SFENCE` to guarantee these writes complete before new writes are handled. At this point, the previous writes that were flushed in the RDMA Send list are now persistent.

Performance Implications of the General-Purpose Remote Replication Method

The general-purpose remote replication method requires that RDMA of the initiator software follows a number of RDMA Write(s) with an RDMA Send. After the target NIC finishes flushing the requested regions, an RDMA Send from the target goes back to the initiator to affirm that the initiator application can consider those writes persistent. This additional send/receive/send/receive messaging has an effect on latency and throughput to make the writes persistent and has 50% higher latency than the appliance remote replication method. The extra messaging has an effect on overall bandwidth and scalability of all the RDMA connections running on those NICs.

Also, if the size of the RDMA Write that needs to be made persistent is small, the efficiency of the connection drops dramatically as the extra messaging overhead becomes a significant component of the overall latency. Additionally, the target node CPU and caches are consumed for that operation. The same data is essentially transmitted twice: once from NIC (via PCIe) to the CPU L3 cache and then from the CPU L3 cache to the memory controller (iMC).

Appliance Remote Replication Method

Users of persistent memory on an Intel platform can use non-allocating write flows by enabling the feature on the specific PCI root complex where incoming writes from the NIC will enter into the CPU's internal fabric and out to the persistent memory. Using the non-allocating write flow, the incoming RDMA Writes will bypass CPU caches and go directly to the persistence domain. This means that writes do not need to be flushed to the persistence domain by the target system CPU.

The I/O pipeline still needs to be flushed to the persistence domain. This is more efficiently accomplished by issuing a small RDMA Read to any memory address on the same RDMA connection as the RDMA Writes; the memory address does not need to be one that was written or is persistent. The RDMA specification clearly states that an RDMA Read will force the previous RDMA Writes to complete first. This ordering rule is also true of the PCIe interconnect to which the target NIC is connected. PCIe Reads will perform a pipeline flush and force previous PCIe writes to complete first.

Figure 18-3 outlines the basic appliance remote replication method, often referred to as the appliance persistency method, described earlier.

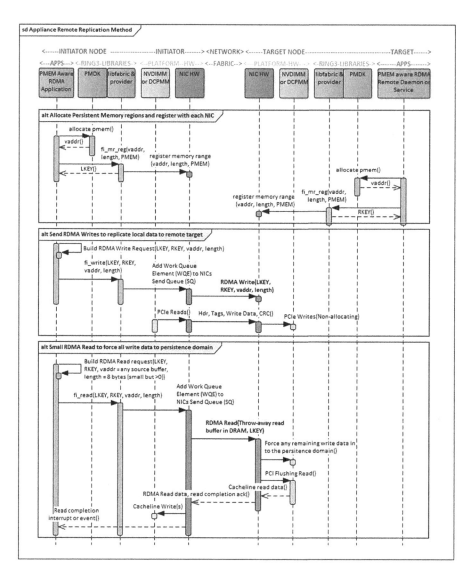

Figure 18-3. *The appliance remote replication method*

How Does the Appliance Remote Replication Method Make Data Persistent?

The combination of bypassing CPU caches on the target system for the inbound RDMA Writes to persistent memory with the ordering semantics of the RDMA and PCIe protocols results in an efficient mechanism to make data persistent. Since the RDMA Read to persistent memory will force previous writes first to persistent memory and the persistence domain, the RDMA Read completion that comes back after those writes are complete is the initiator application's acknowledgment that those writes are now durable.

Chapter 2 defines the persistence domain in depth, including how the platform ensures that all writes get to the media from the persistence domain in the event of a power loss.

Performance Implications of the Appliance Remote Replication Method

This single extra round trip using an RDMA Read is roughly 50% lower latency than the general-purpose server persistency method, which requires two round-trip messages before the writes can be declared durable. As with the first method, as the size of the writes to be made durable gets smaller, the RDMA Read round-trip overhead becomes a significant component of the overall latency.

General Software Architecture

The software stack for the use of remote persistent memory typically uses the same memory-mapped files discussed in Chapter 3. Persistent memory is presented to the RDMA application as a memory-mapped file. The application registers the persistent memory with the local NIC on both ends of the connection, and the resulting registry key is shared with the initiator application for use in the RDMA Read and Write requests. This is the identical process required for using traditional volatile DRAM with RDMA.

A layering of kernel and application-level software components is typically used to allow an application to make use of both persistent memory and an RDMA connection. The IBTA defines verbs interfaces that are typically implemented by the kernel drivers for the NIC and the middleware software application library. Additional libraries may be layered above the verbs layer to provide generic RDMA services via a common API- and NIC-specific provider that implements the library.

On Linux, the Open Fabric Alliance (OFA) libibverbs library provides ring-3 interfaces to configure and use the RDMA connection for NICs that support IB, RoCE, and iWARP RDMA network protocols. The OFA libfabric ring-3 application library can be layered on top of libibverbs to provide a generic high-level common API that can be used with typical RDMA NICs. This common API requires a provider plug-in to implement the common API for the specific network protocol. The OFA web site contains many example applications and performance tests that can be used on Linux with a variety of RDMA-capable NICs. Those examples provide the backbone of the PMDK librpmem library.

Windows implements remotely mounted NTFS volumes via the ring-3 SMB Direct Application library, which provides a number of storage protocols including block storage over RDMA.

Figure 18-4 provides the basic high-level architecture for a typical RDMA application on Linux, using all of the publically available libraries and interfaces. Notice that a separate side-band connection is typically needed to set up the RDMA connections themselves.

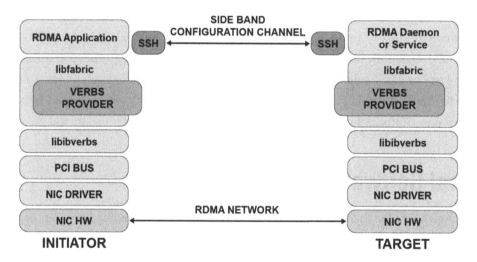

Figure 18-4. *General RDMA software architecture*

librpmem Architecture and Its Use in Replication

PMDK implements both the general-purpose remote replication method and the appliance remote replication method in the librpmem library. As of PMDK v1.7, the librpmem library implements the synchronous and asynchronous replication of local writes to persistent memory on remote systems. librpmem is a low-level library, like libpmem, which allows other libraries to use its replication features.

libpmemobj uses a synchronous write model, meaning that the local initiator write and all of the remotely replicated writes must complete before the local write will be completed back to the application. The libpmemobj library also implements a simple active-passive replication architecture, where all persistent memory transactions are driven through the active initiator node and the remote targets passively standby, replicating the write data. While the passive target systems have the latest write data replicated, the implementation makes no attempt to fail over, fail back, or load balance using the remote systems. The following sections describe the significant performance drawbacks to this implementation.

libpmemobj uses the local memory pool configuration information provided in a configuration file to describe the remote network–connected memory-mapped files. A remote rpmemd program installed on each remote target system is started and connected to the librpmem library on the initiator using a secure encrypted socket connection. Through this connection, librpmem, on behalf of libpmemobj, will set up the RDMA point-to-point connection with each target system, determine the persistence method the target supports (general purpose or appliance method), allocate remote memory-mapped persistent memory files, register the persistent memory on the remote NIC, and retrieve the resulting memory keys for the registered memory.

Once all the RDMA connections to all the targets are established, all required queues are instantiated, and memory buffers have all been allocated and registered, the libpmemobj library is ready to begin remotely replicating all application write data to its local memory-mapped file. When the application calls pmemobj_persist() in libpmemobj, the library will generate a corresponding rpmem_persist() call into librpmem which, in turn, calls the libfabric fi_write() to do the RDMA Write. librpmem then initiates the RDMA Read or Send persistence method (as governed by an understanding of the currently enabled target node's current configuration) by calling libfabric fi_read() or fi_send(). RDMA Read is used in the appliance remote replication method, and RDMA Send is used in the general-purpose remote replication method.

Figure 18-5 outlines the high-level components and interfaces described earlier and used by both the initiator and remote target system using librpmem and libpmemobj.

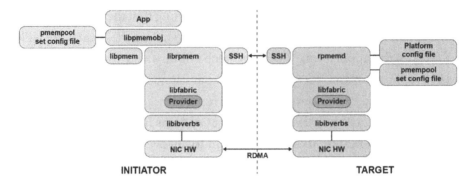

Figure 18-5. *RDMA architecture using libpmemobj and librpmem*

The major components (shown in Figure 18-5) are described in the following to help you understand the high-level architecture that is used by the PMDK's remote replication feature:

librpmem – PMDK Remote RDMA Access Library: The container for the initiator node for all the initiator PMDK functionality that is related to remote replication using RDMA.

rpmemd – PMDK Remote RDMA Configuration Daemon: The container for the target node for all the target PMDK functionality that is related to remote replication using RDMA. It will block any local access to the pmempool set that has been configured for remote usage and executes the remote target interrupt handlers required for the general-purpose remote replication method.

Initiator and Target SSH: This component is used by both librpmem and rpmemd libraries to set up a simple socket connection, close a previously opened socket connection, and send communication packets back and forth.

Libfabric: The OFA defined high-level ring-3 application API for setting up and using a fabric connection in a fabric and vendor-agnostic way. This high-level interface supports RoCE, InfiniBand, and iWARP, as well Intel Omni-Path Architecture products and other network protocols using libfabric-specific transport providers.

Libibverbs: The OFA defined high-level RDMA fabric-based interface. This high-level interface supports RoCE, InfiniBand, and iWARP and is commonly used in most Linux distributions.

Target Node Platform Configuration File: Simple text file generated by the IT admin or user to describe the platform capabilities of the remote target node. This file describes specific capabilities that affect what durability method can be used, that is, ADR-enabled platform, non-allocating write flows enabled by the NIC, and platform type. It also specifies the default socket-connection port that rpmemd will listen on.

Initiator Node PMDK pmempool Set Configuration File: An existing persistent memory poolset configuration file is generated by the system or application administrator that describes local sets of files that will be treated as a pool of persistent memory on the local platform. It also describes local files for local replication and remote target hostnames for remote replication.

Target Node PMDK pmempool Set Configuration File: An existing persistent memory poolset configuration file is generated by the system or application administrator that describes local sets of files that will be treated as a pool of persistent memory on the local platform. On the target node, this set is the collection of files that the initiator node is replicating data into.

Initiator and Target Node Operating System syslog: The standard Linux syslog on each node used by librpmem and rpmemd for outputting useful data for both debug and non-debug information. Since there is little information from rpmemd that is visible on the initiator system, extensive information will be output to the target system syslog when rpmemd is started with the "-d" (debug) runtime option. Even without the debug enabled, rpmemd will output socket events like open, close, create, lost connection, and similar RDMA events.

Configuring Remote Replication Using Poolsets

You are probably already familiar with using poolsets (introduced in Chapter 7) `libpmemobj` to initialize remote replication, which requires two such poolset files. The file used on the initiator side by the `libpmemobj`-enabled application must describe the local memory pool and point to poolset configuration file on the target node, whereas the poolset file on the target node must describe the memory pool shared by the target system.

Listing 18-1 shows a poolset file that will allow replicating local writes to the "remotepool.set" on a remote host.

Listing 18-1. poolwithremotereplica.set – An example of replicating local data to a remote host

```
PMEMPOOLSET
256G /mnt/pmem0/pool1

REPLICA user@example.com remotepool.set
```

Listing 18-2 shows a poolset file that describes the memory-mapped files shared for the remote access. In many ways, a remote poolset file is the same as the regular poolset file, but it must fulfill additional requirements:

- Exist in a poolset directory specified in the `rpmemd` configuration file

- Should be uniquely identified by its name, which an `rpmem`-enabled application has to use to replicate to the specified memory pool

- Cannot define any additional replicas, local or remote

Listing 18-2. remotereplica.set – An example of how to describe the memory pool on the remote host

```
PMEMPOOLSET
256G /mnt/pmem1/pool2
```

Performance Considerations

Once persistent memory is accessible via a remote network connection, significantly lower latency can be achieved compared with writing to a remote SSD or legacy block storage device. This is because the RDMA hardware is writing the remote write data directly into the final persistent memory location, whereas remote replication to an SSD requires an RDMA Write into the DRAM on the remote server, followed by a second local DMA operation to move the remote write data from volatile DRAM into the final storage location on the SSD or other legacy block storage device.

The performance challenge with replicating data to remote persistent memory is that while large block sizes of 512KiB or larger can achieve good performance, as the size of the writes being replicated gets smaller, the network overhead becomes a larger portion of the total latency, and performance can suffer.

If the persistent memory is being used as an SSD replacement, the typical native block storage size is 4K, avoiding some of the inefficiencies seen with small transfers. If the persistent memory replaces a traditional SSD and data is written remotely to the SSD, the latency improvements with persistent memory can be 10x or more.

The synchronous replication model implemented in `librpmem` means that small data structures and pointer updates in local persistent memory result in small, very inefficient, RDMA Writes followed by a small RDMA Read or Send to make that small amount of write data persistent. This results in significant performance degradation compared to writing only to local persistent memory. It makes the replication performance very dependent on the local persistent memory write sequences, which is heavily dependent on the application workload. In general, the larger the average request size and the lower the number of `rpmem_persist()` calls that are required for a given workload will improve the overall latency required for guaranteeing that data is persistent.

It is possible to follow multiple RDMA Writes with single RDMA Read or Send to make all of the preceding writes persistent. This reduces the impact of the size of RDMA Writes on the overall performance of the proposed solution. But using this mitigation, remember you are not guaranteed that any of the RDMA Writes is persistent until RDMA Read completion returns or you receive RDMA Send with a confirmation. The implementation that allows this approach is implemented in `rpmem_flush()` and `rpmem_drain()` API call pair, where `rpmem_flush()` performs RDMA Write and returns immediately and `rpmem_drain()` posts RDMA Read and waits for its completion (at the time of publication it is not implemented in the write/send model).

There are many performance considerations, including the high-level networking model being used. Traditional best-in-class networking architecture typically relies on a *pull* model between the initiator and target. In a pull model, the initiator requests resources from the target, but the target server only pulls the data across via RDMA

Read when it has the resources and connection bandwidth. This server-centric view allows the target node to handle hundreds or thousands of connections since it is in complete control of all resources for all of the connections and initiates the networking transactions when it chooses. With the speed and low latency of persistent memory, a *push* model can be used where the initiator and target have pre-allocated and registered memory resources and directly RDMA Write the data without waiting for server-side resource coordination. Microsoft's SNIA DevCon RDMA presentation describes the push/pull model in more detail: (`https://www.snia.org/sites/default/files/ SDC/2018/presentations/PM/Talpey_Tom_Remote_Persistent_Memory.pdf`).

Remote Replication Error Handling

`librpmem` replication failures will occur for either a lost socket connection or a lost RDMA connection. Any error status returned from `rpmem_persist()`, `rpmem_flush()`, and `rpmem_drain()` is typically treated as an unrecoverable failure. The `libpmemobj` user of `librpmem` API should treat this as a lost socket or RDMA condition and should wait for all remaining `librpmem` API calls to complete, call `rpmem_close()` to close the connection and clean up the stack, and then force the application to exit. When the application restarts, the files will be reopened on both ends, and `libpmemobj` will check only the file metadata. We recommend you do not proceed before synchronizing local and remote memory pools with the pmempool-sync(1) command.

Say Hello to the Replicated World

The beauty of the `libpmemobj` remote replication is that it does not require any changes to the existing `libpmemobj` application. If you take any `libpmemobj` application and provide it with the poolset file configured to use the remote replica, it will simply start replicating. No coding required.

To illustrate how to replicate persistent memory, we look at a Hello World type program demonstrating the replication process directly using the `librpmem` library. Listing 18-3 shows a part of the C program that writes the "Hello world" message to remote memory. If it discovers that the message in English is already there, it translates it to Spanish and writes it back to remote memory. We walk through the lines of the program at the end of the listing.

Listing 18-3. The main routine of the Hello World program with replication

```
37    #include <assert.h>
38    #include <errno.h>
39    #include <unistd.h>
40    #include <stdio.h>
41    #include <stdlib.h>
42    #include <string.h>
43
44    #include <librpmem.h>
45
46    /*
47     * English and Spanish translation of the message
48     */
49    enum lang_t {en, es};
50    static const char *hello_str[] = {
51        [en] = "Hello world!",
52        [es] = "¡Hola Mundo!"
53    };
54
55    /*
56     * structure to store the current message
57     */
58    #define STR_SIZE    100
59    struct hello_t {
60        enum lang_t lang;
61        char str[STR_SIZE];
62    };
63
64    /*
65     * write_hello_str -- write a message to the local memory
66     */
```

```
 67    static inline void
 68    write_hello_str(struct hello_t *hello, enum lang_t lang)
 69    {
 70        hello->lang = lang;
 71        strncpy(hello->str, hello_str[hello->lang], STR_SIZE);
 72    }

104    int
105    main(int argc, char *argv[])
106    {
107        /* for this example, assume 32MiB pool */
108        size_t pool_size = 32 * 1024 * 1024;
109        void *pool = NULL;
110        int created;
111
112        /* allocate a page size aligned local memory pool */
113        long pagesize = sysconf(_SC_PAGESIZE);
114        assert(pagesize >= 0);
115        int ret = posix_memalign(&pool, pagesize, pool_size);
116        assert(ret == 0 && pool != NULL);
117
118        /* skip to the beginning of the message */
119        size_t hello_off = 4096; /* rpmem header size */
120        struct hello_t *hello = (struct hello_t *)(pool + hello_off);
121
122        RPMEMpool *rpp = remote_open("target", "pool.set", pool,
           pool_size,
123                &created);
124        if (created) {
125            /* reset local memory pool */
126            memset(pool, 0, pool_size);
127            write_hello_str(hello, en);
128        } else {
129            /* read message from the remote pool */
130            ret = rpmem_read(rpp, hello, hello_off, sizeof(*hello), 0);
131            assert(ret == 0);
```

```
132
133            /* translate the message */
134            const int lang_num = (sizeof(hello_str) / sizeof(hello_
               str[0]));
135            enum lang_t lang = (enum lang_t)((hello->lang + 1) %
               lang_num);
136            write_hello_str(hello, lang);
137        }
138
139        /* write message to the remote pool */
140        ret = rpmem_persist(rpp, hello_off, sizeof(*hello), 0, 0);
141        printf("%s\n", hello->str);
142        assert(ret == 0);
143
144        /* close the remote pool */
145        ret = rpmem_close(rpp);
146        assert(ret == 0);
147
148        /* release local memory pool */
149        free(pool);
150        return 0;
151    }
```

- Line 68: Simple helper routine for writing message to the local memory.

- Line 115: Allocate a big enough block of memory, which is aligned to the page size. The required block size is hard-coded, whereas the alignment is required if you want to make this memory block available for RDMA transfers.

- Line 122: The remote_open() routine creates or opens the remote memory pool.

- Lines 126-127: The local memory pool is initialized here. It is performed only once when the remote memory pool was just created, so it does not contain any message.

- Line 130: A message from the remote memory pool is read to the local memory here.

- Lines 134-136: If a message from the remote memory pool was read correctly, it is translated locally.

- Line 140: The newly initialized or translated message is written to the remote memory pool.

- Line 145: Close the remote memory pool.

- Line 149: Release remote memory pool.

The last missing piece of the whole process is how the remote replication is set up. It is all done in the remote_open() routine presented in Listing 18-4.

Listing 18-4. A remote_open routine from the Hello World program with replication

```
74    /*
75     * remote_open -- setup the librpmem replication
76     */
77    static inline RPMEMpool*
78    remote_open(const char *target, const char *poolset, void *pool,
79            size_t pool_size, int *created)
80    {
81        /* fill pool_attributes */
82        struct rpmem_pool_attr pool_attr;
83        memset(&pool_attr, 0, sizeof(pool_attr));
84        strncpy(pool_attr.signature, "HELLO", RPMEM_POOL_HDR_SIG_LEN);
85
86        /* create a remote pool */
87        unsigned nlanes = 1;
88        RPMEMpool *rpp = rpmem_create(target, poolset, pool, pool_
size, &nlanes,
89                &pool_attr);
90        if (rpp) {
91            *created = 1;
92            return rpp;
93        }
94
```

```
95        /* create failed so open a remote pool */
96        assert(errno == EEXIST);
97        rpp = rpmem_open(target, poolset, pool, pool_size, &nlanes,
          &pool_attr);
98        assert(rpp != NULL);
99        *created = 0;
100
101       return rpp;
102   }
```

- Line 88: A remote memory pool can be either created or opened.
  When it is used for the first time, it must be created so that it is
  available for the opening afterward. We first try to create it here.

- Line 97: Here we attempt to open the remote memory pool. We
  assume it exists because of the error code received during the create
  try (EEXIST).

Execution Example

The Hello World application produces the output shown in Listing 18-5.

Listing 18-5. An output from the Hello World application for librpmem

```
[user@initiator]$ ./hello
Hello world!
[user@initiator]$ ./hello
¡Hola Mundo!
```

Listing 18-6 shows the contents of the target persistent memory pool where we see
the "Hola Mundo" string.

Listing 18-6. The ¡Hola Mundo! snooped on the replication target

```
[user@target]$ hexdump -s 4096 -C /mnt/pmem1/pool2
00001000  01 00 00 00 c2 a1 48 6f  6c 61 20 4d 75 6e 64 6f  |......Hola
Mundo|
00001010  21 00 00 00 00 00 00 00  00 00 00 00 00 00 00
00  |!..............|
```

```
00001020   00 00 00 00 00 00 00 00  00 00 00 00 00 00 00
00  |...............|
*
00002000
```

Summary

It is important to know that neither the general-purpose remote replication method nor the appliance remote replication method is ideal because vendor-specific platform features are required to use non-allocating writes, adding the complication of effecting performance on an entire PCI root complex. Conversely, flushing remote writes using allocating writes requires a painful interrupt of the target system to intercept an RDMA Send request and flush the list of regions contained within the send buffer. Waking the remote node is extremely painful in a cloud environment because there are hundreds or thousands of inbound RDMA requests from many different connections; avoid this if possible.

There are cloud service providers using these two methods today and getting phenomenal performance results. If the persistent memory is used as a replacement for a remotely accessed SSD, huge reductions in latency can be achieved.

As the first iteration of remote persistence support, we focused on application/ library changes to implement these high-level persistence methods, without hardware, firmware, driver, or protocol changes. At the time of publication, IBTA and IETF drafts for a new wire protocol extension for persistent memory is nearing completion. This will provide native hardware support for RDMA to persistent memory and allow hardware entities to route each I/ O to its destination memory device without the need to change allocating write mode and without the potential to adversely affect performance on collateral devices connected to the same root port. See Appendix E for more details on the new extensions to RDMA, specifically for remote persistence.

RDMA protocol extensions are only one step into further remote persistent memory development. Several other areas of improvement are already identified and shall be addressed to the remote persistent memory users community, including atomicity of remote operations, advanced error handling (including RAS), dynamic configuration of remote persistent memory and custom setup, and real 0% CPU utilization on remote/ target replication side.

As this book has demonstrated, unlocking the true potential of persistent memory may require new approaches to existing software and application architecture. Hopefully, this chapter gave you an overview of this complex topic, the challenges of working with remote persistent memory, and the many aspects of software architecture to consider when unlocking the true performance potential.

## CHAPTER 19

Advanced Topics

This chapter covers several topics that we briefly described earlier in the book but did not expand upon as it would have distracted from the focus points. The in-depth details on these topics are here for your reference.

Nonuniform Memory Access (NUMA)

NUMA is a computer memory design used in multiprocessing where the memory access time depends on the memory location relative to the processor. NUMA is used in a symmetric multiprocessing (SMP) system. An SMP system is a "tightly coupled and share everything" system in which multiple processors working under a single operating system can access each other's memory over a common bus or "interconnect" path. With NUMA, a processor can access its own local memory faster than nonlocal memory (memory that is local to another processor or memory shared between processors). The benefits of NUMA are limited to particular workloads, notably on servers where the data is often associated strongly with certain tasks or users.

CPU memory access is always fastest when the CPU can access its local memory. Typically, the CPU socket and the closest memory banks define a NUMA node. Whenever a CPU needs to access the memory of another NUMA node, it cannot access it directly but is required to access it through the CPU owning the memory. Figure 19-1 shows a two-socket system with DRAM and persistent memory represented as "memory."

© The Author(s) 2020
S. Scargall, *Programming Persistent Memory*, https://doi.org/10.1007/978-1-4842-4932-1_19

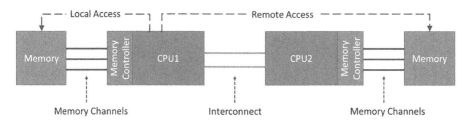

Figure 19-1. *A two-socket CPU NUMA architecture showing local and remote memory access*

On a NUMA system, the greater the distance between a processor and a memory bank, the slower the processor's access to that memory bank. Performance-sensitive applications should therefore be configured so they allocate memory from the closest possible memory bank.

Performance-sensitive applications should also be configured to execute on a set number of cores, particularly in the case of multithreaded applications. Because first-level caches are usually small, if multiple threads execute on one core, each thread will potentially evict cached data accessed by a previous thread. When the operating system attempts to multitask between these threads, and the threads continue to evict each other's cached data, a large percentage of their execution time is spent on cache line replacement. This issue is referred to as cache thrashing. We therefore recommend that you bind a multithreaded application to a NUMA node rather than a single core, since this allows the threads to share cache lines on multiple levels (first-, second-, and last-level cache) and minimizes the need for cache fill operations. However, binding an application to a single core may be performant if all threads are accessing the same cached data. numactl allows you to bind an application to a particular core or NUMA node and to allocate the memory associated with a core or set of cores to that application.

NUMACTL Linux Utility

On Linux we can use the numactl utility to display the NUMA hardware configuration and control which cores and threads application processes can run. The libnuma library included in the numactl package offers a simple programming interface to the NUMA policy supported by the kernel. It is useful for more fine-grained tuning than the numactl utility. Further information is available in the numa(7) man page.

The numactl --hardware command displays an inventory of the available NUMA nodes within the system. The output shows only volatile memory, not persistent memory. We will show how to use the ndctl command to show NUMA locality of persistent memory in the next section. The number of NUMA nodes does not always equal the number of sockets. For example, an AMD Threadripper 1950X has 1 socket and 2 NUMA nodes. The following output from numactl was collected from a two-socket Intel Xeon Platinum 8260L processor server with a total of 385GiB DDR4, 196GiB per socket.

```
# numactl --hardware
available: 2 nodes (0-1)
node 0 cpus: 0 1 2 3 4 5 6 7 8 9 10 11 12 13 14 15 16 17 18 19 20 21 22 23
48 49 50 51 52 53 54 55 56 57 58 59 60 61 62 63 64 65 66 67 68 69 70 71
node 0 size: 192129 MB
node 0 free: 187094 MB
node 1 cpus: 24 25 26 27 28 29 30 31 32 33 34 35 36 37 38 39 40 41 42 43 44
45 46 47 72 73 74 75 76 77 78 79 80 81 82 83 84 85 86 87 88 89 90 91 92 93
94 95
node 1 size: 192013 MB
node 1 free: 191478 MB
node distances:
node   0   1
  0:  10  21
  1:  21  10
```

The node distance is a relative distance and not an actual time-based latency in nanoseconds or milliseconds.

numactl lets you bind an application to a particular core or NUMA node and allocate the memory associated with a core or set of cores to that application. Some useful options provided by numactl are described in Table 19-1.

Table 19-1. *numactl command options for binding processes to NUMA nodes or CPUs*

Option	Description
--membind, -m	Only allocate memory from specific NUMA nodes. The allocation will fail if there is not enough memory available on these nodes.
--cpunodebind, -N	Only execute the process on CPUs from the specified NUMA nodes.
--physcpubind, -C	Only execute process on the given CPUs.
--localalloc, -l	Always allocate on the current NUMA node.
--preferred	Preferably allocate memory on the specified NUMA node. If memory cannot be allocated, fall back to other nodes.

NDCTL Linux Utility

The ndctl utility is used to create persistent memory capacity for the operating system, called namespaces, as well as enumerating, enabling, and disabling the dimms, regions, and namespaces. Using the –v (verbose) option shows what NUMA node (numa_node) persistent memory DIMMS (-D), regions (-R), and namespaces (-N) belong to. Listing 19-1 shows the region and namespaces for a two-socket system. We can correlate the numa_node with the corresponding NUMA node shown by the numactl command.

Listing 19-1. Region and namespaces for a two-socket system

```
# ndctl list -Rv
{
  "regions":[
    {
      "dev":"region1",
      "size":1623497637888,
      "available_size":0,
      "max_available_extent":0,
      "type":"pmem",
      "numa_node":1,
```

```
    "iset_id":-2506113243053544244,
    "persistence_domain":"memory_controller",
    "namespaces":[
      {
        "dev":"namespace1.0",
        "mode":"fsdax",
        "map":"dev",
        "size":1598128390144,
        "uuid":"b3e203a0-2b3f-4e27-9837-a88803f71860",
        "raw_uuid":"bd8abb69-dd9b-44b7-959f-79e8cf964941",
        "sector_size":512,
        "align":2097152,
        "blockdev":"pmem1",
        "numa_node":1
      }
    ]
  },
  {
    "dev":"region0",
    "size":1623497637888,
    "available_size":0,
    "max_available_extent":0,
    "type":"pmem",
    "numa_node":0,
    "iset_id":3259620181632232652,
    "persistence_domain":"memory_controller",
    "namespaces":[
      {
        "dev":"namespace0.0",
        "mode":"fsdax",
        "map":"dev",
        "size":1598128390144,
        "uuid":"06b8536d-4713-487d-891d-795956d94cc9",
        "raw_uuid":"39f4abba-5ca7-445b-ad99-fd777f7923c1",
        "sector_size":512,
        "align":2097152,
```

```
            "blockdev":"pmem0",
            "numa_node":0
        }
      ]
    }
  ]
}
```

Intel Memory Latency Checker Utility

To get absolute latency numbers between NUMA nodes on Intel systems, you can use the Intel Memory Latency Checker (Intel MLC), available from `https://software.intel.com/en-us/articles/intel-memory-latency-checker`.

Intel MLC provides several modes specified through command-line arguments:

- `--latency_matrix` prints a matrix of local and cross-socket memory latencies.

- `--bandwidth_matrix` prints a matrix of local and cross-socket memory bandwidths.

- `--peak_injection_bandwidth` prints peak memory bandwidths of the platform for various read-write ratios.

- `--idle_latency` prints the idle memory latency of the platform.

- `--loaded_latency` prints the loaded memory latency of the platform.

- `--c2c_latency` prints the cache-to-cache data transfer latency of the platform.

Executing `mlc` or `mlc_avx512` with no arguments runs all the modes in sequence using the default parameters and values for each test and writes the results to the terminal. The following example shows running just the latency matrix on a two-socket Intel system.

```
# ./mlc_avx512 --latency_matrix -e -r
Intel(R) Memory Latency Checker - v3.6
Command line parameters: --latency_matrix -e -r
```

```
Using buffer size of 2000.000MiB
Measuring idle latencies (in ns)...
                Numa node
Numa node            0       1
        0         84.2   141.4
        1        141.5    82.4
```

- `--latency_matrix` prints a matrix of local and cross-socket memory latencies.

- `-e` means that the hardware prefetcher states do not get modified.

- `-r` is random access reads for latency thread.

MLC can be used to test persistent memory latency and bandwidth in either DAX or FSDAX modes. Commonly used arguments include

- `-L` requests that large pages (2MB) be used (assuming they have been enabled).

- `-h` requests huge pages (1GB) for DAX file mapping.

- `-J` specifies a directory in which files for `mmap` will be created (by default no files are created). This option is mutually exclusive with `-j`.

- `-P CLFLUSH` is used to evict stores to persistent memory.

Examples:

Sequential read latency:

```
# mlc_avx512 --idle_latency -J/mnt/pmemfs
```

Random read latency:

```
# mlc_avx512 --idle_latency -l256 -J/mnt/pmemfs
```

NUMASTAT Utility

The numastat utility on Linux shows per NUMA node memory statistics for processors and the operating system. With no command options or arguments, it displays NUMA hit and miss system statistics from the kernel memory allocator. The default numastat statistics shows per-node numbers, in units of pages of memory, for example:

```
$ sudo numastat
                        node0            node1
numa_hit             8718076          7881244
numa_miss                  0                0
numa_foreign               0                0
interleave_hit         40135            40160
local_node           8642532          2806430
other_node             75544          5074814
```

- numa_hit is memory successfully allocated on this node as intended.

- numa_miss is memory allocated on this node despite the process preferring some different node. Each numa_miss has a numa_foreign on another node.

- numa_foreign is memory intended for this node but is actually allocated on a different node. Each numa_foreign has a numa_miss on another node.

- interleave_hit is interleaved memory successfully allocated on this node as intended.

- local_node is memory allocated on this node while a process was running on it.

- other_node is memory allocated on this node while a process was running on another node.

Intel VTune Profiler – Platform Profiler

On Intel systems, you can use the Intel VTune Profiler - Platform Profiler, previously called VTune Amplifier, (discussed in Chapter 15) to show CPU and memory statistics, including hit and miss rates of CPU caches and data accesses to DDR and persistent memory. It can also depict the system's configuration to show what memory devices are physically located on which CPU.

IPMCTL Utility

Persistent memory vendor- and server-specific utilities can also be used to show DDR and persistent memory device topology to help identify what devices are associated with which CPU sockets. For example, the `ipmctl show -topology` command displays the DDR and persistent memory (non-volatile) devices with their physical memory slot location (see Figure 19-2), if that data is available.

```
$ sudo ipmctl show -topology

DimmID | MemoryType                    | Capacity  | PhysicalID| DeviceLocat
================================================================================
0x0001 | Logical Non-Volatile Device   | 252.4 GiB | 0x0028    | CPU1_DIMM_A2
0x0011 | Logical Non-Volatile Device   | 252.4 GiB | 0x002c    | CPU1_DIMM_B2
0x0021 | Logical Non-Volatile Device   | 252.4 GiB | 0x0030    | CPU1_DIMM_C2
0x0101 | Logical Non-Volatile Device   | 252.4 GiB | 0x0036    | CPU1_DIMM_D2
0x0111 | Logical Non-Volatile Device   | 252.4 GiB | 0x003a    | CPU1_DIMM_E2
0x0121 | Logical Non-Volatile Device   | 252.4 GiB | 0x003e    | CPU1_DIMM_F2
0x1001 | Logical Non-Volatile Device   | 252.4 GiB | 0x0044    | CPU2_DIMM_A2
0x1011 | Logical Non-Volatile Device   | 252.4 GiB | 0x0048    | CPU2_DIMM_B2
0x1021 | Logical Non-Volatile Device   | 252.4 GiB | 0x004c    | CPU2_DIMM_C2
0x1101 | Logical Non-Volatile Device   | 252.4 GiB | 0x0052    | CPU2_DIMM_D2
0x1111 | Logical Non-Volatile Device   | 252.4 GiB | 0x0056    | CPU2_DIMM_E2
0x1121 | Logical Non-Volatile Device   | 252.4 GiB | 0x005a    | CPU2_DIMM_F2
N/A    | DDR4                          | 32.0 GiB  | 0x0026    | CPU1_DIMM_A1
N/A    | DDR4                          | 32.0 GiB  | 0x002a    | CPU1_DIMM_B1
N/A    | DDR4                          | 32.0 GiB  | 0x002e    | CPU1_DIMM_C1
N/A    | DDR4                          | 32.0 GiB  | 0x0034    | CPU1_DIMM_D1
N/A    | DDR4                          | 32.0 GiB  | 0x0038    | CPU1_DIMM_E1
N/A    | DDR4                          | 32.0 GiB  | 0x003c    | CPU1_DIMM_F1
N/A    | DDR4                          | 32.0 GiB  | 0x0042    | CPU2_DIMM_A1
N/A    | DDR4                          | 32.0 GiB  | 0x0046    | CPU2_DIMM_B1
N/A    | DDR4                          | 32.0 GiB  | 0x004a    | CPU2_DIMM_C1
N/A    | DDR4                          | 32.0 GiB  | 0x0050    | CPU2_DIMM_D1
N/A    | DDR4                          | 32.0 GiB  | 0x0054    | CPU2_DIMM_E1
N/A    | DDR4                          | 32.0 GiB  | 0x0058    | CPU2_DIMM_F1
```

Figure 19-2. *Topology report from the ipmctl show –topology command*

BIOS Tuning Options

The BIOS contains many tuning options that change the behavior of CPU, memory, persistent memory, and NUMA. The location and name may vary between server platform types, server vendors, persistent memory vendors, or BIOS versions. However, most applicable tunable options can usually be found in the Advanced menu under Memory Configuration and Processor Configuration. Refer to your system BIOS user manual for descriptions of each available option. You may want to test several BIOS options with the application(s) to understand which options bring the most value.

Automatic NUMA Balancing

Physical limitations to hardware are encountered when many CPUs and a lot of memory are required. The important limitation is the limited communication bandwidth between the CPUs and the memory. The NUMA architecture modification addresses this issue. An application generally performs best when the threads of its processes are accessing memory on the same NUMA node as the threads are scheduled. Automatic NUMA balancing moves tasks (which can be threads or processes) closer to the memory they are accessing. It also moves application data to memory closer to the tasks that reference it. The kernel does this automatically when automatic NUMA balancing is active. Most operating systems implement this feature. This section discusses the feature on Linux; refer to your Linux distribution documentation for specific options as they may vary.

Automatic NUMA balancing is enabled by default in most Linux distributions and will automatically activate at boot time when the operating system detects it is running on hardware with NUMA properties. To determine if the feature is enabled, use the following command:

```
$ sudo cat /proc/sys/kernel/numa_balancing
```

A value of 1 (true) indicates the feature is enabled, whereas a value of 0 (zero/false) means it is disabled.

Automatic NUMA balancing uses several algorithms and data structures, which are only active and allocated if automatic NUMA balancing is active on the system, using a few simple steps:

- A task scanner periodically scans the address space and marks the memory to force a page fault when the data is next accessed.

- The next access to the data will result in a NUMA Hinting Fault. Based on this fault, the data can be migrated to a memory node associated with the thread or process accessing the memory.

- To keep a thread or process, the CPU it is using and the memory it is accessing together, the scheduler groups tasks that share data.

Manual NUMA tuning of applications using `numactl` will override any system-wide automatic NUMA balancing settings. Automatic NUMA balancing simplifies tuning workloads for high performance on NUMA machines. Where possible, we recommend statically tuning the workload to partition it within each node. Certain latency-sensitive applications, such as databases, usually work best with manual configuration. However, in most other use cases, automatic NUMA balancing should help performance.

Using Volume Managers with Persistent Memory

We can provision persistent memory as a block device on which a file system can be created. Applications can access persistent memory using standard file APIs or memory map a file from the file system and access the persistent memory directly through load/store operations. The accessibility options are described in Chapters 2 and 3.

The main advantages of volume managers are increased abstraction, flexibility, and control. Logical volumes can have meaningful names like "databases" or "web." Volumes can be resized dynamically as space requirements change and migrated between physical devices within the volume group on a running system.

On NUMA systems, there is a locality factor between the CPU and the DRR and persistent memory that is directly attached to it. Accessing memory on a different CPU across the interconnect incurs a small latency penalty. Latency-sensitive applications, such as databases, understand this and coordinate their threads to run on the same socket as the memory they are accessing.

Compared with SSD or NVMe capacity, persistent memory is relatively small. If your application requires a single file system that consumes all persistent memory on the system rather than one file system per NUMA node, a software volume manager can be used to create concatenations or stripes (RAID0) using all the system's capacity. For example, if you have 1.5TiB of persistent memory per CPU socket on a two-socket system, you could build a concatenation or stripe (RAID0) to create a 3TiB file system. If local system redundancy is more important than large file systems, mirroring (RAID1) persistent memory across NUMA nodes is possible. In general, replicating the data across physical servers for redundancy is better. Chapter 18 discusses remote persistent memory in detail, including using remote direct memory access (RDMA) for data transfer and replication across systems.

There are too many volume manager products to provide step-by-step recipes for all of them within this book. On Linux, you can use Device Mapper (dmsetup), Multiple Device Driver (mdadm), and Linux Volume Manager (LVM) to create volumes that use the capacity from multiple NUMA nodes. Because most modern Linux distributions default to using LVM for their boot disks, we assume that you have some experience using LVM. There is extensive information and tutorials within the Linux documentation and on the Web.

Figure 19-3 shows two regions on which we can create either an fsdax or sector type namespace that creates the corresponding /dev/pmem0 and /dev/pmem1 devices. Using /dev/pmem[01], we can create an LVM physical volume which we then combine to create a volume group. Within the volume group, we are free to create as many logical volumes of the requested size as needed. Each logical volume can support one or more file systems.

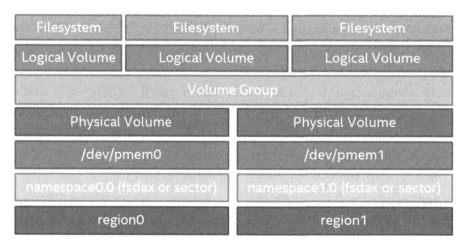

Figure 19-3. *Linux Volume Manager architecture using persistent memory regions and namespaces*

We can also create a number of possible configurations if we were to create multiple namespaces per region or partition the /dev/pmem* devices using fdisk or parted, for example. Doing this provides greater flexibility and isolation of the resulting logical volumes. However, if a physical NVDIMM fails, the impact is significantly greater since it would impact some or all of the file systems depending on the configuration.

Creating complex RAID volume groups may protect the data but at the cost of not efficiently using all the persistent memory capacity for data. Additionally, complex RAID volume groups do not support the DAX feature that some applications may require.

The mmap( ) MAP_SYNC Flag

Introduced in the Linux kernel v4.15, the MAP_SYNC flag ensures that any needed file system metadata writes are completed before a process is allowed to modify directly mapped data. The MAP_SYNC flag was added to the mmap( ) system call to request the synchronous behavior; in particular, the guarantee provided by this flag is

While a block is writeably mapped into page tables of this mapping, it is guaranteed to be visible in the file at that offset also after a crash.

This means the file system will not silently relocate the block, and it will ensure that the file's metadata is in a consistent state so that the blocks in question will be present after a crash. This is done by ensuring that any needed metadata writes were done before the process is allowed to write pages affected by that metadata.

When a persistent memory region is mapped using MAP_SYNC, the memory management code will check to see whether there are metadata writes pending for the affected file. However, it will not actually flush those writes out. Instead, the pages are mapped read only with a special flag, forcing a page fault when the process first attempts to perform a write to one of those pages. The fault handler will then synchronously flush out any dirty metadata, set the page permissions to allow the write, and return. At that point, the process can write the page safely, since all the necessary metadata changes have already made it to persistent storage.

The result is a relatively simple mechanism that will perform far better than the currently available alternative of manually calling fsync( ) before each write to persistent memory. The additional IO from fsync( ) can potentially cause the process to block in what was supposed to be a simple memory write, introducing latency that may be unexpected and unwanted.

The mmap(2) man page in the Linux Programmer's manual describes the MAP_SYNC flag as follows:

MAP_SYNC (since Linux 4.15)

This flag is available only with the MAP_SHARED_VALIDATE mapping type; mappings of type MAP_SHARED will silently ignore this flag. This flag is supported only for files supporting DAX (direct mapping of persistent memory). For other files, creating a mapping with this flag results in an EOPNOTSUPP error.

Shared file mappings with this flag provide the guarantee that while some memory is writably mapped in the address space of the process, it will be visible in the same file at the same offset even after the system crashes or is rebooted. In conjunction with the use of appropriate CPU instructions, this provides users of such mappings with a more efficient way of making data modifications persistent.

Summary

In this chapter, we presented some of the more advanced topics for persistent memory including page size considerations on large memory systems, NUMA awareness and how it affects application performance, how to use volume managers to create DAX file systems that span multiple NUMA nodes, and the MAP_SYNC flag for mmap(). Additional topics such as BIOS tuning were intentionally left out of this book as it is vendor and product specific. Performance and benchmarking of persistent memory products are left to external resources as there are too many tools – vdbench, sysbench, fio, etc. – and too many options for each one, to cover in this book.

APPENDIX A

How to Install NDCTL and DAXCTL on Linux

The `ndctl` utility is used to manage the libnvdimm (non-volatile memory device) subsystem in the Linux kernel and to administer namespaces. The `daxctl` utility provides enumeration and provisioning commands for any device-dax namespaces you create. `daxctl` is only required if you work directly with device-dax namespaces. We presented a use-case for the 'system-ram' dax type in Chapter 10, that can use persistent memory capacity to dynamically extend the usable volatile memory capacity in Linux. Chapter 10 also showed how libmemkind can use device dax namespaces for volatile memory in addition to using DRAM. The default, and recommended, namespace for most developers is filesystem-dax (fsdax). Both Linux-only utilities - `ndctl` and `daxctl` - are open source and are intended to be persistent memory vendor neutral. Microsoft Windows has integrated graphical utilities and PowerShell Commandlets to administer persistent memory.

libndctl and libdaxctl are required for several Persistent Memory Development Kit (PMDK) features if compiling from source. If ndctl is not available, the PMDK may not build all components and features, but it will still successfully compile and install. In this appendix, we describe how to install `ndctl` and `daxctl` using the Linux package repository only. To compile ndctl from source code, refer to the README on the ndctl GitHub repository (`https://github.com/pmem/ndctl`) or `https://docs.pmem.io`.

Prerequisites

Installing ndctl and daxctl using packages automatically installs any missing dependency packages on the system. A full list of dependencies is usually listed when installing the package. You can query the package repository to list dependencies or use an online package took such as `https://pkgs.org` to find the package for your operating

© The Author(s) 2020
S. Scargall, *Programming Persistent Memory*, https://doi.org/10.1007/978-1-4842-4932-1

system and list the package details. For example, Figure A-1 shows the packages required for ndctl v64.1 on Fedora 30 (`https://fedora.pkgs.org/30/fedora-x86_64/ndctl-64.1-1.fc30.x86_64.rpm.html`).

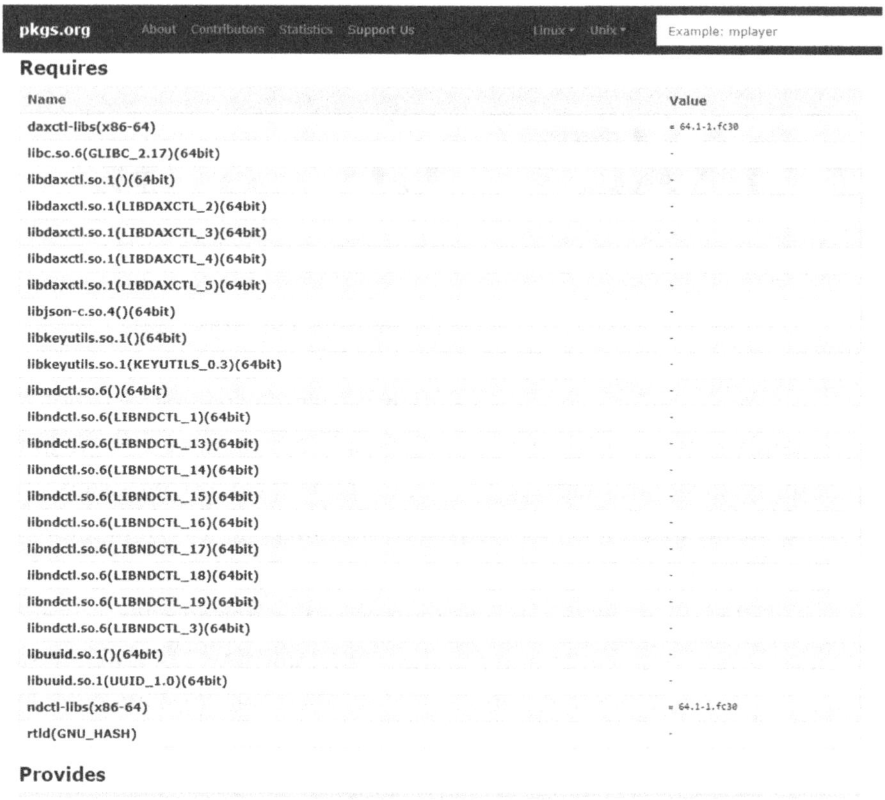

Requires

Name	Value
daxctl-libs(x86-64)	= 64.1-1.fc30
libc.so.6(GLIBC_2.17)(64bit)	-
libdaxctl.so.1()(64bit)	-
libdaxctl.so.1(LIBDAXCTL_2)(64bit)	-
libdaxctl.so.1(LIBDAXCTL_3)(64bit)	-
libdaxctl.so.1(LIBDAXCTL_4)(64bit)	-
libdaxctl.so.1(LIBDAXCTL_5)(64bit)	-
libjson-c.so.4()(64bit)	-
libkeyutils.so.1()(64bit)	-
libkeyutils.so.1(KEYUTILS_0.3)(64bit)	-
libndctl.so.6()(64bit)	-
libndctl.so.6(LIBNDCTL_1)(64bit)	-
libndctl.so.6(LIBNDCTL_13)(64bit)	-
libndctl.so.6(LIBNDCTL_14)(64bit)	-
libndctl.so.6(LIBNDCTL_15)(64bit)	-
libndctl.so.6(LIBNDCTL_16)(64bit)	-
libndctl.so.6(LIBNDCTL_17)(64bit)	-
libndctl.so.6(LIBNDCTL_18)(64bit)	-
libndctl.so.6(LIBNDCTL_19)(64bit)	-
libndctl.so.6(LIBNDCTL_3)(64bit)	-
libuuid.so.1()(64bit)	-
libuuid.so.1(UUID_1.0)(64bit)	-
ndctl-libs(x86-64)	= 64.1-1.fc30
rtld(GNU_HASH)	-

Provides

***Figure A-1.** Detailed package information for ndctl v64.1 on Fedora 30*

Installing NDCTL and DAXCTL Using the Linux Distribution Package Repository

The ndctl and daxctl utilities are delivered as runtime binaries with the option to install development header files which can be used to integrate their features in to your application or when compiling PMDK from source code. To create debug binaries, you need to compile ndctl and daxctl from source code. Refer to the README on the project page `https://github.com/pmem/ndctl` or `https://docs.pmem.io` for detailed instructions.

Searching for Packages Within a Package Repository

The default package manager utility for your operating system will allow you to query the package repository using regular expressions to identify packages to install. Table A-1 shows how to search the package repository using the command-line utility for several distributions. If you prefer to use a GUI, feel free to use your favorite desktop utility to perform the same search and install operations described here.

Table A-1. *Searching for ndctl and daxctl packages in different Linux distributions*

Operating System	Command
Fedora 21 or Earlier	`$ yum search ndctl` `$ yum search daxctl`
Fedora 22 or Later	`$ dnf search ndctl` `$ dnf search daxctl`
RHEL AND CENTOS	`$ yum search ndctl` `$ yum search daxctl`
SLES AND OPENSUSE	`$ zipper search ndctl` `$ zipper search daxctl`

CANONICAL/Ubuntu	`$ aptitude search ndctl`
	`$ apt-cache search ndctl`
	`$ apt search ndctl`
	`$ aptitude search daxctl`
	`$ apt-cache search daxctl`
	`$ apt search daxctl`

Additionally, you can use an online package search tools such as `https://pkgs.org` that allow you to search for packages across multiple distros. Figure A-2 shows the results for many distros when searching for "libpmem."

Figure A-2. *https://pkgs.org search results for "ndctl"*

Installing NDCTL and DAXCTL from the Package Repository

Instructions for some popular Linux distributions follow. Skip to the section for your operating system. If your operating system is not listed here, it may share the same package family as one listed here so you can use the same instructions. Should your operating system not meet either criteria, see the ndctl project home page https://github.com/pmem/ndctl or https://docs.pmem.io for installation instructions.

> **Note**   The version of the ndctl and daxctl available with your operating system may not match the most current project release. If you require a newer release than your operating system delivers, consider compiling the projects from the source code. We do not describe compiling and installing from the source code in this book. Instructions can be found on `https://docs.pmem.io/getting-started-guide/installing-ndctl#installing-ndctl-from-source-on-linux` and `https://github.com/pmem/ndctl`.

Installing PMDK on Fedora 22 or Later

To install individual packages, you can execute

```
$ sudo dnf install <package>
```

For example, to install just the ndctl runtime utility and library, use

```
$ sudo dnf install ndctl
```

To install all packages, use

```
Runtime:
$ sudo dnf install ndctl daxctl
```

```
Development library:
$ sudo dnf install ndctl-devel
```

Installing PMDK on RHEL and CentOS 7.5 or Later

To install individual packages, you can execute

```
$ sudo yum install <package>
```

For example, to install just the ndctl runtime utility and library, use

```
$ sudo yum install ndctl
```

To install all packages, use

Runtime:
```
$ yum install ndctl daxctl
```

Development:
```
$ yum install ndctl-devel
```

Installing PMDK on SLES 12 and OpenSUSE or Later

To install individual packages, you can execute

```
$ sudo zypper install <package>
```

For example, to install just the ndctl runtime utility and library, use

```
$ sudo zypper install ndctl
```

To install all packages, use

All Runtime:
```
$ zypper install ndctl daxctl
```

All Development:
```
$ zypper install libndctl-devel
```

Installing PMDK on Ubuntu 18.04 or Later

To install individual packages, you can execute

```
$ sudo zypper install <package>
```

For example, to install just the ndctl runtime utility and library, use

```
$ sudo zypper install ndctl
```

To install all packages, use

All Runtime:
```
$ sudo apt-get install ndctl daxctl
```

All Development:
```
$ sudo apt-get install libndctl-dev
```

APPENDIX B

How to Install the Persistent Memory Development Kit (PMDK)

The Persistent Memory Development Kit (PMDK) is available on supported operating systems in package and source code formats. Some features of the PMDK require additional packages. We describe instructions for Linux and Windows.

PMDK Prerequisites

In this appendix, we describe installing the PMDK libraries using the packages available in your operating system package repository. To enable all PMDK features, such as advanced reliability, accessibility, and serviceability (RAS), PMDK requires libndctl and libdaxctl. Package dependencies automatically install these requirements. If you are building and installing using the source code, you should install NDCTL first using the instructions provided in Appendix C.

Installing PMDK Using the Linux Distribution Package Repository

The PMDK is a collection of different libraries; each one provides different functionality. This provides greater flexibility for developers as only the required runtime or header files need to be installed without installing unnecessary libraries.

© The Author(s) 2020
S. Scargall, *Programming Persistent Memory*, https://doi.org/10.1007/978-1-4842-4932-1

Package Naming Convention

Libraries are available in runtime, development header files (*-devel), and debug (*-debug) versions. Table B-1 shows the runtime (libpmem), debug (libpmem-debug), and development and header files (libpmem-devel) for Fedora. Package names may differ between Linux distributions. We provide instructions for some of the common Linux distributions later in this section.

Table B-1. *Example runtime, debug, and development package naming convention*

Library	Description
LIBPMEM	Low-level persistent memory support library
LIBPMEM-DEBUG	Debug variant of the libpmem low-level persistent memory library
LIBPMEM-DEVEL	Development files for the low-level persistent memory library

Searching for Packages Within a Package Repository

Table B-2 shows the list of available libraries as of PMDK v1.6. For an up-to-date list, see `https://pmem.io/pmdk`.

Table B-2. *PMDK libraries as of PMDK v1.6*

Library	Description
LIBPMEM	Low-level persistent memory support library
LIBRPMEM	Remote Access to persistent memory library
LIBPMEMBLK	Persistent Memory Resident Array of Blocks library
LIBPMEMCTO	Close-to-Open Persistence library (Deprecated in PMDK v1.5)
LIBPMEMLOG	Persistent Memory Resident Log File library
LIBPMEMOBJ	Persistent Memory Transactional Object Store library
LIBPMEMPOOL	Persistent Memory pool management library
PMEMPOOL	Utilities for Persistent Memory

The default package manager utility for your operating system will allow you to query the package repository using regular expressions to identify packages to install. Table B-3 shows how to search the package repository using the command-line utility for several distributions. If you prefer to use a GUI, feel free to use your favorite desktop utility to perform the same search and install operations described here.

Table B-3. *Searching for *pmem* packages on different Linux operating systems*

Operating System	Command
Fedora 21 or Earlier	`$ yum search pmem`
Fedora 22 or Later	`$ dnf search pmem`
	`$ dnf repoquery *pmem*`
RHEL AND CENTOS	`$ yum search pmem`
SLES AND OPENSUSE	`$ zipper search pmem`
CANONICAL/Ubuntu	`$ aptitude search pmem`
	`$ apt-cache search pmem`
	`$ apt search pmem`

Additionally, you can use an online package search tools such as `https://pkgs.org` that allow you to search for packages across multiple distros. Figure B-1 shows the results for many distros when searching for "libpmem."

Figure B-1. *Search results for "libpmem" on* `https://pkgs.org`

Installing PMDK Libraries from the Package Repository

Instructions for some popular Linux distributions follow. Skip to the section for your operating system. If your operating system is not listed here, it may share the same package family as one listed here so you can use the same instructions. Should your operating system not meet either criteria, see `https://docs.pmem.io` for installation instructions and the PMDK project home page (`https://github.com/pmem/pmdk`) to see the most recent instructions.

Note   The version of the PMDK libraries available with your operating system may not match the most current PMDK release. If you require a newer release than your operating system delivers, consider compiling PMDK from the source code. We do not describe compiling and installing PMDK from the source code in this book. Instructions can be found on `https://docs.pmem.io/getting-started-guide/installing-pmdk/compiling-pmdk-from-source` and `https://github.com/pmem/pmdk`.

Installing PMDK on Fedora 22 or Later

To install individual libraries, you can execute

```
$ sudo dnf install <library>
```

For example, to install just the libpmem runtime library, use

```
$ sudo dnf install libpmem
```

To install all packages, use

All Runtime:
```
$ sudo dnf install libpmem librpmem libpmemblk libpmemlog/
  libpmemobj libpmempool pmempool
```

All Development:
```
$ sudo dnf install libpmem-devel librpmem-devel \
  libpmemblk-devel libpmemlog-devel libpmemobj-devel \
  libpmemobj++-devel libpmempool-devel
```

All Debug:
```
$ sudo dnf install libpmem-debug librpmem-debug \
  libpmemblk-debug libpmemlog-debug libpmemobj-debug \
  libpmempool-debug
```

Installing PMDK on RHEL and CentOS 7.5 or Later

To install individual libraries, you can execute

```
$ sudo yum install <library>
```

For example, to install just the libpmem runtime library, use

```
$ sudo yum install libpmem
```

To install all packages, use

All Runtime:
```
$ sudo yum install libpmem librpmem libpmemblk libpmemlog \
    libpmemobj libpmempool pmempool
```

All Development:
```
$ sudo yum install libpmem-devel librpmem-devel \
    libpmemblk-devel libpmemlog-devel libpmemobj-devel \
    libpmemobj++-devel libpmempool-devel
```

All Debug:
```
$ sudo yum install libpmem-debug librpmem-debug \
    libpmemblk-debug libpmemlog-debug libpmemobj-debug \
    libpmempool-debug
```

Installing PMDK on SLES 12 and OpenSUSE or Later

To install individual libraries, you can execute

```
$ sudo zypper install <library>
```

For example, to install just the libpmem runtime library, use

```
$ sudo zypper install libpmem
```

To install all packages, use

All Runtime:
```
$ sudo zypper install libpmem librpmem libpmemblk libpmemlog \
    libpmemobj libpmempool pmempool
```

All Development:
```
$ sudo zypper install libpmem-devel librpmem-devel \
    libpmemblk-devel libpmemlog-devel libpmemobj-devel \
    libpmemobj++-devel libpmempool-devel
```

All Debug:
```
$ sudo zypper install libpmem-debug librpmem-debug \
    libpmemblk-debug libpmemlog-debug libpmemobj-debug \
    libpmempool-debug
```

Installing PMDK on Ubuntu 18.04 or Later

To install individual libraries, you can execute

```
$ sudo zypper install <library>
```

For example, to install just the libpmem runtime library, use

```
$ sudo zypper install libpmem
```

To install all packages, use

All Runtime:
```
$ sudo apt-get install libpmem1 librpmem1 libpmemblk1 \
    libpmemlog1 libpmemobj1 libpmempool1
```

All Development:
```
$ sudo apt-get install libpmem-dev librpmem-dev \
    libpmemblk-dev libpmemlog-dev libpmemobj-dev \
    libpmempool-dev libpmempool-dev
```

All Debug:
```
$ sudo apt-get install libpmem1-debug \
    librpmem1-debug libpmemblk1-debug \
    libpmemlog1-debug libpmemobj1-debug libpmempool1-debug
```

Installing PMDK on Microsoft Windows

The recommended and easiest way to install PMDK on Windows is to use Microsoft vcpkg. Vcpkg is an open source tool and ecosystem created for library management. To build PMDK from source that can be used in a different packaging or development solution, see the README on `https://github.com/pmem/pmdk` or `https://docs.pmem.io`.

To install the latest PMDK release and link it to your Visual Studio solution, you first need to clone and set up vcpkg on your machine as described on the vcpkg GitHub page (`https://github.com/Microsoft/vcpkg`).

In brief:

```
> git clone https://github.com/Microsoft/vcpkg
> cd vcpkg
> .\bootstrap-vcpkg.bat
> .\vcpkg integrate install
> .\vcpkg install pmdk:x64-windows
```

Note   The last command can take a while as PMDK builds and installs.

After successful completion of all of the preceding steps, the libraries are ready to be used in Visual Studio with no additional configuration is required. Just open Visual Studio with your existing project or create a new one (remember to use platform **x64**) and then include headers to project as you always do.

APPENDIX C

How to Install IPMCTL on Linux and Windows

The `ipmctl` utility is used to configure and manage Intel Optane DC persistent memory modules (DCPMM). This is a vendor-specific utility available for Linux and Windows. It supports functionality to:

- Discover DCPMMs on the platform

- Provision the platform memory configuration

- View and update the firmware on DCPMMs

- Configure data-at-rest security on DCPMMs

- Monitor DCPMM health

- Track performance of DCPMMs

- Debug and troubleshoot DCPMMs

`ipmctl` refers to the following interface components:

- `libipmctl`: An application programming interface (API) library for managing PMMs

- `ipmctl`: A command-line interface (CLI) application for configuring and managing PMMs from the command line

- `ipmctl-monitor`: A monitor daemon/system service for monitoring the health and status of PMMs

© The Author(s) 2020
S. Scargall, *Programming Persistent Memory*, https://doi.org/10.1007/978-1-4842-4932-1

IPMCTL Linux Prerequisites

ipmctl requires libsafec as a dependency.

libsafec

libsafec is available as a package in the Fedora package repository. For other Linux distributions, it is available as a separate downloadable package for local installation:

- RHEL/CentOS EPEL 7 packages can be found at https://copr.fedorainfracloud.org/coprs/jhli/safeclib/.

- OpenSUSE/SLES packages can be found at https://build.opensuse.org/package/show/home:jhli/safeclib.

- Ubuntu packages can be found at https://launchpad.net/~jhli/+archive/ubuntu/libsafec.

Alternately, when compiling ipmctl from source code, use the -DSAFECLIB_SRC_DOWNLOAD_AND_STATIC_LINK=ON option to download sources and statically link to safeclib.

IPMCTL Linux Packages

As a vendor-specific utility, it is not included in most Linux distribution package repositories other than Fedora. EPEL7 packages can be found at https://copr.fedorainfracloud.org/coprs/jhli/ipmctl. OpenSUSE and SLES packages can be found at https://build.opensuse.org/package/show/home:jhli/ipmctl.

IPMCTL for Microsoft Windows

The latest Windows EXE binary for ipmctl can be downloaded from the "Releases" section of the GitHub project page (https://github.com/intel/ipmctl/releases) as shown in Figure C-1.

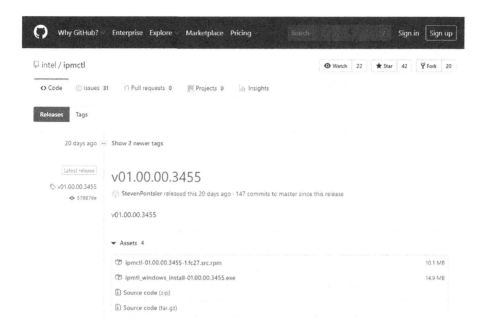

Figure C-1. *ipmctl releases on GitHub (`https://github.com/intel/ipmctl/releases`)*

Running the executable installs ipmctl and makes it available via the command-line and PowerShell interfaces.

Using ipmctl

The ipmctl utility provides system administrators with the ability to configure Intel Optane DC persistent memory modules which can then be used by Windows PowerShellCmdlets or ndctl on Linux to create namespaces on which file systems can be created. Applications can then create persistent memory pools and memory map them to get direct access to the persistent memory. Detailed information about the modules can also be extracted to help with errors or debugging.

ipmctl has a rich set of commands and options that can be displayed by running ipmctl without any command verb, as shown in Listing C-1.

Listing C-1. Listing the command verbs and simple usage information

```
# ipmctl version

Intel(R) Optane(TM) DC Persistent Memory Command Line Interface Version
01.00.00.3279

# ipmctl

Intel(R) Optane(TM) DC Persistent Memory Command Line Interface

    Usage: ipmctl <verb>[<options>][<targets>][<properties>]
Commands:
    Display the CLI help.
    help

    Display the CLI version.
    version

    Update the firmware on one or more DIMMs
    load -source (File Source) -dimm[(DimmIDs)]

    Set properties of one/more DIMMs such as device security and modify
    device.
    set -dimm[(DimmIDs)]

    Erase persistent data on one or more DIMMs.
    delete -dimm[(DimmIDs)]

    Show information about one or more Regions.
    show -region[(RegionIDs)] -socket(SocketIDs)

    Provision capacity on one or more DIMMs into regions
    create -dimm[(DimmIDs)] -goal -socket(SocketIDs)

    Show region configuration goal stored on one or more DIMMs
    show -dimm[(DimmIDs)] -goal -socket[(SocketIDs)]

    Delete the region configuration goal from one or more DIMMs
    delete -dimm[(DimmIDs)] -goal -socket(SocketIDs)
```

Load stored configuration goal for specific DIMMs
load -source (File Source) -dimm[(DimmIDs)] -goal -socket(SocketIDs)

Store the region configuration goal from one or more DIMMs to a file
dump -destination (file destination) -system -config

Modify the alarm threshold(s) for one or more DIMMs.
set -sensor(List of Sensors) -dimm[(DimmIDs)]

Starts a playback or record session
start -session -mode -tag

Stops the active playback or recording session.
stop -session

Dump the PBR session buffer to a file
dump -destination (file destination) -session

Show basic information about session pbr file
show -session

Load Recording into memory
load -source (File Source) -session

Clear the namespace LSA partition on one or more DIMMs
delete -dimm[(DimmIDs)] -pcd[(Config)]

Show error log for given DIMM
show -error(Thermal|Media) -dimm[(DimmIDs)]

Dump firmware debug log
dump -destination (file destination) -debug -dimm[(DimmIDs)]

Show information about one or more DIMMs.
show -dimm[(DimmIDs)] -socket[(SocketIDs)]

Show basic information about the physical processors in the host
server.
show -socket[(SocketIDs)]

Show health statistics
show -sensor[(List of Sensors)] -dimm[(DimmIDs)]

Run a diagnostic test on one or more DIMMs
start -diagnostic[(Quick|Config|Security|FW)] -dimm[(DimmIDs)]

Show the topology of the DCPMMs installed in the host server
show -topology -dimm[(DimmIDs)] -socket[(SocketIDs)]

Show information about total DIMM resource allocation.
show -memoryresources

Show information about BIOS memory management capabilities.
show -system -capabilities

Show information about firmware on one or more DIMMs.
show -dimm[(DimmIDs)] -firmware

Show the ACPI tables related to the DIMMs in the system.
show -system[(NFIT|PCAT|PMTT)]

Show pool configuration goal stored on one or more DIMMs
show -dimm[(DimmIDs)] -pcd[(Config|LSA)]

Show user preferences and their current values
show -preferences

Set user preferences
set -preferences

Show Command Access Policy Restrictions for DIMM(s).
show -dimm[(DimmIDs)] -cap

Show basic information about the host server.
show -system -host

Show event stored on one in the system log
show -event -dimm[(DimmIDs)]

Set event's action required flag on/off
set -event(EventID) ActionRequired=(0)

Capture a snapshot of the system state for support purposes
dump -destination (file destination) -support

```
Show performance statistics per DIMM
show -dimm[(DimmIDs)] -performance[(Performance Metrics)]
```

```
Please see ipmctl <verb> -help <command> i.e 'ipmctl show -help -dimm' for
more information on specific command
```

Each command has its own man page. A full list of man pages can be found from the IPMCTL(1) man page by running "man ipmctl".

An online ipmctl User Guide can be found at https://docs.pmem.io. This guide provides detailed step-by-step instructions and in-depth information about ipmctl and how to use it to provision and debug issues. An ipmctl Quick Start Guide can be found at https://software.intel.com/en-us/articles/quick-start-guide-configure-intel-optane-dc-persistent-memory-on-linux.

For a short video walk-through of using ipmctl and ndctl, you can watch the "Provision Intel Optane DC Persistent Memory in Linux" webinar recording (https://software.intel.com/en-us/videos/provisioning-intel-optane-dc-persistent-memory-modules-in-linux).

If you have questions relating to ipmctl, Intel Optane DC persistent memory, or a general persistent memory question, you can ask it in the Persistent Memory Google Forum (https://groups.google.com/forum/#!forum/pmem). Questions or issues specific to ipmctl should be posted as an issue or question on the ipmctl GitHub issues site (https://github.com/intel/ipmctl/issues).

APPENDIX D

Java for Persistent Memory

Java is one of the most popular programming languages available because it is fast, secure, and reliable. There are lots of applications and web sites implemented in Java. It is cross-platform and supports multi-CPU architectures from laptops to datacenters, game consoles to scientific supercomputers, cell phones to the Internet, and CD/DVD players to automotive. Java is everywhere!

At the time of writing this book, Java did not natively support storing data persistently on persistent memory, and there were no Java bindings for the Persistent Memory Development Kit (PMDK), so we decided Java was not worthy of a dedicated chapter. We didn't want to leave Java out of this book given its popularity among developers, so we decided to include information about Java in this appendix.

In this appendix, we describe the features that have already been integrated in to Oracle's Java Development Kit (JDK) [https://www.oracle.com/java/] and OpenJDK [https://openjdk.java.net/]. We also provide information about proposed persistent memory functionality in Java as well as two external Java libraries in development.

Volatile Use of Persistent Memory

Java does support persistent memory for volatile use cases on systems that have heterogeneous memory architectures. That is a system with DRAM, persistent memory, and non-volatile storage such as SSD or NVMe drives.

© The Author(s) 2020
S. Scargall, *Programming Persistent Memory*, https://doi.org/10.1007/978-1-4842-4932-1

Heap Allocation on Alternative Memory Devices

Both Oracle JDK v10 and OpenJDK v10 implemented JEP 316: Heap allocation on alternative memory devices [http://openjdk.java.net/jeps/316]. The goal of this feature is to enable the HotSpot VM to allocate the Java object heap on an alternative memory device, such as persistent memory, specified by the user.

As described in Chapter 3, Linux and Windows can expose persistent memory through the file system. Examples are NTFS and XFS or ext4. Memory-mapped files on these direct access (DAX) file systems bypass the page cache and provide a direct mapping of virtual memory to the physical memory on the device.

To allocate the Java heap using memory-mapped files on a DAX file system, Java added a new runtime option, -XX:AllocateHeapAt=<path>. This option takes a path to the DAX file system and uses memory mapping to allocate the object heap on the memory device. Using this option enables the HotSpot VM to allocate the Java object heap on an alternative memory device, such as persistent memory, specified by the user. The feature does not intend to share a non-volatile region between multiple running JVMs or reuse the same region for further invocations of the JVM.

Figure D-1 shows the architecture of this new heap allocation method using both DRAM and persistent memory backed virtual memory.

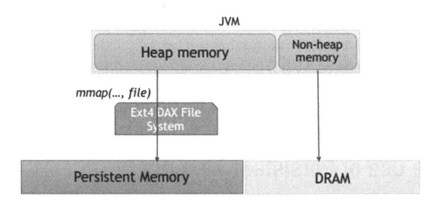

Figure D-1. *Java heap memory allocated from DRAM and persistent memory using the "-XX:AllocateHeapAt=<path>" option*

The Java heap is allocated only from persistent memory. The mapping to DRAM is shown to emphasize that non-heap components like code cache, gc bookkeeping, and so on, are allocated from DRAM.

The existing heap-related flags such as -Xmx, -Xms, and garbage collection–related flags will continue to work as before. For example:

```
$ java –Xmx32g –Xms16g –XX:AllocateHeapAt=/pmemfs/jvmheap \
ApplicationClass
```

This allocates an initial 16GiB heap size (-Xms) with a maximum heap size up to 32GiB (-Xmx32g). The JVM heap can use the capacity of a temporary file created within the path specified by --XX:AllocateHeapAt=/pmemfs/jvmheap. JVM automatically creates a temporary file of the form jvmheap.XXXXX, where XXXXX is a randomly generated number. The directory path should be a persistent memory backed file system mounted with the DAX option. See Chapter 3 for more information about mounting file systems with the DAX feature.

To ensure application security, the implementation must ensure that file(s) created in the file system are:

- Protected by correct permissions, to prevent other users from accessing it

- Removed when the application terminates, in any possible scenario

The temporary file is created with read-write permissions for the user running the JVM, and the JVM deletes the file before terminating.

This feature targets alternative memory devices that have the same semantics as DRAM, including the semantics of atomic operations, and can therefore be used instead of DRAM for the object heap without any change to existing application code. All other memory structures such as the code heap, metaspace, thread stacks, etc., will continue to reside in DRAM.

Some use cases of this feature include

- In multi-JVM deployments, some JVMs such as daemons, services, etc., have lower priority than others. Persistent memory would potentially have higher access latency compared to DRAM. Low-priority processes can use persistent memory for the heap, allowing high-priority processes to use more DRAM.

- Applications such as big data and in-memory databases have an ever-increasing demand for memory. Such applications could use persistent memory for the heap since persistent memory modules would potentially have a larger capacity compared to DRAM.

More information about this feature can be found in these resources:

- Oracle JavaSE 10 Documentation [`https://docs.oracle.com/javase/10/tools/java.htm#GUID-3B1CE181-CD30-4178-9602-230B800D4FAE__BABCBGHF`]

- OpenJDK JEP 316: Heap Allocation on Alternative Memory Devices [`http://openjdk.java.net/jeps/316`]

Partial Heap Allocation on Alternative Memory Devices

HotSpot JVM 12.0.1 introduced a feature to allocate old generation of Java heap on an alternative memory device, such as persistent memory, specified by the user.

The feature in G1 and parallel GC allows them to allocate part of heap memory in persistent memory to be used exclusively for old generation objects. The rest of the heap is mapped to DRAM, and young generation objects are always placed here.

Operating systems expose persistent memory devices through the file system, so the underlying media can be accessed directly, or direct access (DAX). File systems that support DAX include NTFS on Microsoft Windows and ext4 and XFS on Linux. Memory-mapped files in these file systems bypass the file cache and provide a direct mapping of virtual memory to the physical memory on the device. The specification of a path to a DAX mounted file system uses the flag `-XX:AllocateOldGenAt=<path>` which enables this feature. There are no additional flags to enable this feature.

When enabled, young generation objects are placed in DRAM only, while old generation objects are always allocated in persistent memory. At any given point, the garbage collector guarantees that the total memory committed in DRAM and persistent memory is always less than the size of the heap as specified by `-Xmx`.

When enabled, the JVM also limits the maximum size of the young generation based on available DRAM, although it is recommended that users set the maximum size of the young generation explicitly.

For example, if the JVM is executed with `-Xmx756g` on a system with 32GB DRAM and 1024GB persistent memory, the garbage collector will limit the young generation size based on the following rules:

- No `-XX:MaxNewSize` or `-Xmn` is specified: The maximum young generation size is set to 80% of available memory (25.6GB).

- `-XX:MaxNewSize` or `-Xmn` is specified: The maximum young generation size is capped at 80% of available memory (25.6GB) regardless of the amount specified.

- Users can use `-XX:MaxRAM` to let the VM know how much DRAM is available for use. If specified, maximum young gen size is set to 80% of the value in MaxRAM.

- Users can specify the percentage of DRAM to use, instead of the default 80%, for young generation with

- `-XX:MaxRAMPercentage`.

- Enabling logging with the logging option `gc+ergo=info` will print the maximum young generation size at startup.

Non-volatile Mapped Byte Buffers

JEP 352: Non-Volatile Mapped Byte Buffers [`https://openjdk.java.net/jeps/352`] adds a new JDK-specific file mapping mode so that the `FileChannel` API can be used to create `MappedByteBuffer` instances that refer to persistent memory. The feature should be available in Java 14 when it is released, which is after the publication of this book.

This JEP proposes to upgrade `MappedByteBuffer` to support access to persistent memory. The only API change required is a new enumeration employed by `FileChannel` clients to request mapping of a file located on a DAX file system rather than a conventional, file storage system. Recent changes to the `MappedByteBufer` API mean that it supports all the behaviors needed to allow direct memory updates and provide the durability guarantees needed for higher level, Java client libraries to implement persistent data types (e.g., block file systems, journaled logs, persistent objects, etc.). The implementations of `FileChannel` and `MappedByteBuffer` need revising to be aware of this new backing type for the mapped file.

The primary goal of this JEP is to ensure that clients can access and update persistent memory from a Java program efficiently and coherently. A key element of this goal is to ensure that individual writes (or small groups of contiguous writes) to a buffer region can be committed with minimal overhead, that is, to ensure that any changes which might still be in cache are written back to memory.

A second, subordinate goal is to implement this commit behavior using a restricted, JDK-internal API defined in class unsafe, allowing it to be reused by classes other than `MappedByteBuffer` that may need to commit to persistent memory.

A final, related goal is to allow buffers mapped over persistent memory to be tracked by the existing monitoring and management APIs.

It is already possible to map a persistent memory device file to a `MappedByteBuffer` and commit writes using the current `force()` method, for example, using Intel's libpmem library as device driver or by calling out to libpmem as a native library. However, with the current API, both those implementations provide a "sledgehammer" solution. A force cannot discriminate between clean and dirty lines and requires a system call or JNI call to implement each writeback. For both those reasons, the existing capability fails to satisfy the efficiency requirement of this JEP.

The target OS/CPU platform combinations for this JEP are Linux/x64 and Linux/AArch64. This restriction is imposed for two reasons. This feature will only work on OSes that support the `mmap` system call MAP_SYNC flag, which allows synchronous mapping of non-volatile memory. That is true of recent Linux releases. It will also only work on CPUs that support cache line writeback under user space control. x64 and AArch64 both provide instructions meeting this requirement.

Persistent Collections for Java (PCJ)

The Persistent Collections for Java library (PCJ) is an open source Java library being developed by Intel for persistent memory programming. More information on PCJ, including source code and sample code, is available on GitHub at `https://github.com/pmem/pcj`.

At the time of writing this book, the PCJ library was still defined as a "pilot" project and still in an experimental state. It is being made available now in the hope it is useful in exploring the retrofit of existing Java code to use persistent memory as well as exploring persistent Java programming in general.

The library offers a range of thread-safe persistent collection classes including arrays, lists, and maps. It also offers persistent support for things like strings and primitive integer and floating-point types. Developers can define their own persistent classes as well.

Instances of these persistent classes behave much like regular Java objects, but their fields are stored in persistent memory. Like regular Java objects, their lifetime is reachability-based; they are automatically garbage collected if there are no outstanding references to them. Unlike regular Java objects, their lifetime can extend beyond a single instance of the Java virtual machine and beyond machine restarts.

Because the contents of persistent objects are retained, it's important to maintain data consistency of objects even in the face of crashes and power failures. Persistent collections and other objects in the library offer persistent data consistency at the Java method level. Methods, including field setters, behave as if the method's changes to persistent memory all happen or none happen. This same method-level consistency can be achieved with developer-defined classes using a transaction API offer by PCJ.

PCJ uses the libpmemobj library from the Persistent Memory Development Kit (PMDK) which we discussed in Chapter 7. For additional information on PMDK, please visit `https://pmem.io/` and `https://github.com/pmem/pmdk`.

Using PCJ in Java Applications

To import this library into an existing Java application, include the project's target/ classes directory in your Java `classpath` and the project's target/cppbuild directory in your `java.library.path`. For example:

```
$ javac -cp .:<path>/pcj/target/classes <source>
$ java -cp .:<path>/pcj/target/classes \
    -Djava.library.path=<path>/pcj/target/cppbuild <class>
```

There are several ways to use the PCJ library:

1. Use instances of built-in persistent classes in your applications.

2. Extend built-in persistent classes with new methods.

3. Declare new persistent classes or extend built-in classes with methods and persistent fields.

PCJ source code examples can be found in the resources listed in the following:

- Introduction to Persistent Collections for Java – `https://github.com/pmem/pcj/blob/master/Introduction.txt`

- Code Sample: Introduction to Java* API for Persistent Memory Programming – `https://software.intel.com/en-us/articles/code-sample-introduction-to-java-api-for-persistent-memory-programming`

417

- Code Sample: Create a "Hello World" Program Using Persistent Collections for Java∗ (PCJ) – `https://software.intel.com/en-us/articles/code-sample-create-a-hello-world-program-using-persistent-collections-for-java-pcj`

Low-Level Persistent Library (LLPL)

The Low-Level Persistence Library (LLPL) is an open source Java library being developed by Intel for persistent memory programming. By providing Java access to persistent memory at a memory block level, LLPL gives developers a foundation for building custom abstractions or retrofitting existing code. More information on LLPL, including source code, sample code, and javadocs, is available on GitHub at `https://github.com/pmem/llpl`.

The library offers management of heaps of persistent memory and manual allocation and deallocation of blocks of persistent memory within a heap. A Java persistent memory block class provides methods to read and write Java integer types within a block as well as copy bytes from block to block and between blocks and (volatile) Java byte arrays.

Several different kinds of heaps and corresponding memory blocks are available to aid in implementing different data consistency schemes. Examples of such implementable schemes:

- Transactional: Data in memory is usable after a crash or power failure

- Persistent: Data in memory is usable after a controlled process exit

- Volatile: Persistent memory used for its large capacity, data is not needed after exit.

Mixed data consistency schemes are also implementable. For example, transactional writes for critical data and either persistent or volatile writes for less critical data (e.g., statistics or caches).

LLPL uses the libpmemobj library from the Persistent Memory Development Kit (PMDK) which we discussed in Chapter 7. For additional information on PMDK, please visit `https://pmem.io/` and `https://github.com/pmem/pmdk`.

Using LLPL in Java Applications

To use LLPL with your Java application, you need to have PMDK and LLPL installed on your system. To compile the Java classes, you need to specify the LLPL class path. Assuming you have LLPL installed on your home directory, do the following:

```
$ javac -cp .:/home/<username>/llpl/target/classes LlplTest.java
```

After that, you should see the generated *.class file. To run the main() method inside your class, you need to again pass the LLPL class path. You also need to set the java.library.path environment variable to the location of the compiled native library used as a bridge between LLPL and PMDK:

```
$ java -cp .:/.../llpl/target/classes \
-Djava.library.path=/.../llpl/target/cppbuild LlplTest
```

PCJ source code examples can be found in the resources listed in the following:

- Code Sample: Introducing the Low-Level Persistent Library (LLPL) for Java* – https://software.intel.com/en-us/articles/introducing-the-low-level-persistent-library-llpl-for-java

- Code Sample: Create a "Hello World" Program Using the Low-Level Persistence Library (LLPL) for Java* – https://software.intel.com/en-us/articles/code-sample-create-a-hello-world-program-using-the-low-level-persistence-library-llpl-for-java

- Enabling Persistent Memory Use in Java – https://www.snia.org/sites/default/files/PM-Summit/2019/presentations/05-PMSummit19-Dohrmann.pdf

Summary

At the time of writing this book, native support for persistent memory in Java is an ongoing effort. Current features are mostly volatile, meaning the data is not persisted once the app exits. We have described several features that have been integrated and shown two libraries – LLPL and PCJ – that provide additional functionality for Java applications.

The Low-Level Persistent Library (LLPL) is an open source Java library being developed by Intel for persistent memory programming. By providing Java access to persistent memory at a memory block level, LLPL gives developers a foundation for building custom abstractions or retrofitting existing code.

The higher-level Persistent Collections for Java (PCJ) offers developers a range of thread-safe persistent collection classes including arrays, lists, and maps. It also offers persistent support for things like strings and primitive integer and floating-point types. Developers can define their own persistent classes as well.

APPENDIX E

The Future of Remote Persistent Memory Replication

As discussed in Chapter 18, the general purpose and appliance remote persistent memory methods are simple high-level upper-layer-protocol (ULP) changes. These methods add a secondary RDMA Send or RDMA Read after a number of RDMA Writes to remote persistent memory. One of the pain points with these implementations is the Intel-specific platform feature, allocating writes, which, by default, pushes inbound PCIe Write data from the NIC directly into the lowest-level CPU cache, speeding the local software access to that newly written data. For persistent memory, it is desirable to turn off allocating writes to persistent memory, elevating the need to flush the CPU cache to guarantee persistence. However, the platform limitations on the control over allocating writes only imprecise control over the behavior of writes for an entire PCIe Root complex. All devices connected to a given root complex will behave the same way. The implications to other software running on the system can be difficult to determine if access to the write data is delayed by bypassing caches. These are contradictory requirements since allocating writes should be disabled for writes to persistent memory, but for writes to volatile memory, allocating writes should be enabled.

To make this per IO steering possible, the networking hardware and software need to have native support for persistent memory. If the networking stack is aware of the persistent memory regions, it can select whether the write is steered toward the persistent memory subsystem or the volatile memory subsystem on a per IO basis, completely removing the need to change global PCIe Root complex allocating-write settings.

S. Scargall, *Programming Persistent Memory*, https://doi.org/10.1007/978-1-4842-4932-1

Also, if the hardware is aware of writes to persistent memory, some significant performance gains can be seen with certain workloads by the reduction in the number of round trip completions that software must wait for. This pipeline efficiency gains are estimated to yield a 30-50% reduction in round-trip latency for the common database SQL Tail-of-Log use case where a large write to persistent memory is followed by an 8-byte pointer update, to be written only after the first remote write data is considered in the persistence domain. The first-generation software remote persistent methods require two software round-trip completions for the initial SQL data write and again for the small 8-byte pointer update write, as shown in Figure E-1A. In the improved native hardware solution shown in Figure E-1B, software waits for a single round-trip completion across the network.

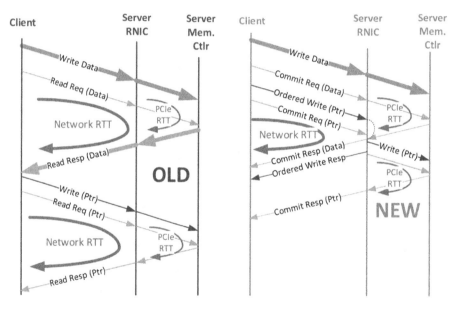

Figure E-1. *The proposed RDMA protocol changes to efficiently support persistent memory by avoiding Send or Read being called after a Write*

These performance improvements are coming in a future Intel platform, native Intel RDMA-capable NICs, and through industry networking standards. Other vendor's RDMA-capable NICs will also support the improved standard. Broad adoption is required to allow users of any vendor's NIC with any vendor's persistent memory on any number of platforms. To accomplish this, native persistent memory support is being driven into the standardized iWarp wire protocol by the IETF, Internet Engineering Taskforce and the standardized InfiniBand and RoCE wire protocol by the IBTA,

InfiniBand Trade Association. Both protocols track each other architecturally and have essentially added an RDMA Flush and RDMA Atomic Write commands to the existing volatile memory support.

RDMA Flush – Is a protocol command that flushes a portion of a memory region. The completion of the flush command indicates that all of the RDMA Writes within the domain of the flush have made it to the final placement. Flush placement hints allow the initiator software to request flushing to globally visible memory (could be volatile or persistent memory regions) and separately whether the memory is volatile or persistent memory. The scope of the RDMA Write data that is included in the RDMA Flush domain is driven by the offset and length for the memory region being flushed. All RDMA Writes covering memory regions contained in the RDMA Flush command shall be included in the RDMA Flush. That means that the RDMA Flush command will not complete on the initiator system until all previous remote writes for those regions have reached the final requested placement location.

RDMA Atomic Write – Is a protocol command that instructs the NIC to write a pointer update directly into persistent memory in a pipeline efficient manner. This allows the preceding RDMA Write, RDMA Flush, RDMA Atomic Write, and RDMA Flush sequence to occur with only one single complete round-trip latency incurred by software. It simply needs to wait for the final RDMA Flush completion.

Platform hardware changes are required to efficiently make use of the new network protocol additions for persistent memory support. The placement hints provided in the RDMA Flush command allows four possible routing combinations:

- Cache Attribute

- No-cache Attribute

- Volatile Destination

- Persistent memory destination

The chipset, CPU, and PCIe root complexes need to understand these placement attributes and steer or route the request to the proper hardware blocks as requested.

On upcoming Intel platforms, the CPU will look at the PCIe TLP Processor Hint fields to allow the NIC to add the steering information to each PCIe packet generated for the inbound RDMA Writes and RDMA Flush. The optional use of this PCIe steering mechanism is defined by the PCIe Firmware Interface in the ACPI specification and allows NIC kernel drivers and PCI bus drivers to enable the IO steering and essentially select cache, no-cache as memory attributes, and persistent memory or DRAM as the destination.

From a software enabling point of view, there will be changes to the *verbs* definition as defined by the IBTA. This will define the specifics of how the NIC will manage and implement the feature. Middleware, including OFA `libibverbs` and `libfabric`, will be updated based on these core additions to the networking protocol for native persistent memory support.

Readers seeking more specific information on the development of these persistent memory extensions to RDMA are encouraged to follow the references in this book and the information shared here to begin a more detailed search on native persistent memory support for high-performance remote access. There are many new exciting developments occurring on this aspect of persistent memory usage.

Glossary

Term	Definition
3D XPoint	3D Xpoint is a non-volatile memory (NVM) technology developed jointly by Intel and Micron Technology.
ACPI	The Advanced Configuration and Power Interface is used by BIOS to expose platform capabilities.
ADR	Asynchronous DRAM Refresh is a feature supported on Intel that triggers a flush of write pending queues in the memory controller on power failure. Note that ADR does not flush the processor cache.
AMD	Advanced Micro Devices https://www.amd.com
BIOS	Basic Input/Output System refers to the firmware used to initialize a server.
CPU	Central processing unit
DCPM	Intel Optane DC persistent memory
DCPMM	Intel Optane DC persistent memory module(s)
DDR	Double Data Rate is an advanced version of SDRAM, a type of computer memory.
DDIO	Direct Data IO. Intel DDIO makes the processor cache the primary destination and source of I/O data rather than main memory. By avoiding system memory, Intel DDIO reduces latency, increases system I/O bandwidth, and reduces power consumption due to memory reads and writes.
DRAM	Dynamic random-access memory
eADR	Enhanced Asynchronous DRAM Refresh, a superset of ADR that also flushes the CPU caches on power failure.

(continued)

© The Author(s) 2020
S. Scargall, *Programming Persistent Memory*, https://doi.org/10.1007/978-1-4842-4932-1

Term	Definition
ECC	Memory error correction used to provide protection from both transient errors and device failures.
HDD	A hard disk drive is a traditional spinning hard drive.
InfiniBand	InfiniBand (IB) is a computer networking communications standard used in high-performance computing that features very high throughput and very low latency. It is used for data interconnect both among and within computers. InfiniBand is also used as either a direct or switched interconnect between servers and storage systems, as well as an interconnect between storage systems.
Intel	Intel Corporation `https://intel.com`
iWARP	Internet Wide Area RDMA Protocol is a computer networking protocol that implements remote direct memory access (RDMA) for efficient data transfer over Internet Protocol networks.
NUMA	Nonuniform memory access, a platform where the time to access memory depends on its location relative to the processor.
NVDIMM	A non-volatile dual inline memory module is a type of random-access memory for computers. Non-volatile memory is memory that retains its contents even when electrical power is removed, for example, from an unexpected power loss, system crash, or normal shutdown.
NVMe	Non-volatile memory express is a specification for directly connecting SSDs on PCIe that provides lower latency and higher performance than SAS and SATA.
ODM	Original Design Manufacturing refers to a producer/reseller relationship in which the full specifications of a project are determined by the reseller rather than based on the specs established by the manufacturer.
OEM	An original equipment manufacturer is a company that produces parts and equipment that may be marketed by another manufacturer.
OS	Operating system
PCIe	Peripheral Component Interconnect Express is a high-speed serial communication bus.
Persistent Memory	Persistent memory (PM or PMEM) provides persistent storage of data, is byte addressable, and has near-memory speeds.

(continued)

Term	Definition
PMoF	Persistent memory over fabric
PSU	Power supply unit
RDMA	Remote direct memory access is a direct memory access from the memory of one computer into that of another without involving the operating system.
RoCE	RDMA over Converged Ethernet is a network protocol that allows remote direct memory access (RDMA) over an Ethernet network.
QPI	Intel QuickPath Interconnect is used for multi-socket communication between CPUs.
SCM	Storage class memory, a synonym for persistent memory.
SSD	Solid-state disk drive is a high-performance storage device built using non-volatile memory.
TDP	A thermal design point specifies the amount of power that the CPU can consume and therefore the amount of heat that the platform must be able to remove in order to avoid thermal throttling conditions.
UMA	Uniform memory access, a platform where the timne to access memory is (roughly) the same, regardless of which processor is doing the access. On Intel patforms, this is achieved by interleaving the memory across sockets.

Index

A

ACPI specification, 28
Address range scrub (ARS), 338
Address space layout randomization
(ASLR), 87, 112, 316
Appliance remote replication
method, 355, 357
Application binary interface (ABI), 122
Application startup and recovery
ACPI specification, 28
ARS, 29
dirty shutdown, 27
flow, 27, 28
infinite loop, 28
libpmem library, 27
libpmemobj query, 27
PMDK, 29
RAS, 27
Asynchronous DRAM
Refresh (ADR), 17, 207
Atomicity, consistency, isolation, and
durability (ACID), 278
Atomic operations, 285, 286

B

Block Translation Table (BTT)
driver, 34
Buffer-based LRU design, 182

C

C++ Standard limitations
object layout, 122, 123
object lifetime, 119, 120
vs. persistent memory, 125, 126
pointers, 123–125
trivial types, 120–122
type traits, 125
Cache flush operation (CLWB), 24, 59, 286
Cache hierarchy
CPU
cache hit, 15
cache miss, 16
levels, 14, 15
and memory controller, 14, 15
non-volatile storage devices, 16
Cache thrashing, 374
Chunks/buckets, 188
CLFLUSHOPT, 18, 19, 24, 208, 247, 353
close() method, 151
closeTable() method, 268
CLWB flushing instructions, 208
cmap engine, 4
Concurrent data structures
definition, 287
erase operation, 293
find operation, 292
hash map, 291, 292
insert operation, 292, 293

© The Author(s) 2020
S. Scargall, *Programming Persistent Memory*, https://doi.org/10.1007/978-1-4842-4932-1

Q

Queue implementation, 126, 128
QuickPath Interconnect (QPI)/Ultra Path
Interconnect (UPI), 305

R

RDMA networking protocols
commands, 350
NIC, 349
RDMA Read, 350
RDMA Send (and Receive), 350, 351
RDMA Write, 350
RDMA over Converged Ethernet
(RoCE), 348
Redis, 143
Redo logging, 320
Reliability, Availability, and
Serviceability (RAS)
device health
ACPI NFIT, 343
ACPI specification, 342
unsafe/dirty shutdown, 343
using ndctl to query, 340, 341
vendors, 342
ECC
inifinite loop, 334
MCE, 334
using Linux, 335, 336
unconsumed uncorrectable error
handling
ARS, 338
clearing errors, 339
memory root-device notification, 338
petrol scrub, 337
runtime, 337
unsafe/dirty shutdown,
DLC counter, 344, 345

Reliability, availability,
serviceability (RAS), 27
Remote direct memory access
(RDMA), 12, 347, 348
software architecture, 357, 358
remote_open routine, 368, 369
remove() method, 150
Resource acquisition is initialization
(RAII), 113
rpmem_drain(), 363
rpmem_flush(), 363

S

show() method, 131
Single instruction, multiple data (SIMD)
processing, 190
Single linked list (SLL), 266
Snapshotting optimization, 196
SNIA NVM programming
model, 351
Solid-state disk (SSD), 1, 156
Stack/buffer overflow bug, 208
Stackoverflow app, 211
Standard Template Library
(STL), 168, 282, 287
std::is_standard_layout, 122, 123
Storage and Networking Industry
Association (SNIA), 33, 112
symmetric multiprocessing (SMP)
system, 373

T

Thread migration, 310
Transactions and multithreading
counter, 281
illustrative execution, 281

Transactions and multithreading (*cont.*)
    incorrect synchronization, 279, 280
    libpmemobj, 280
    PMDK transactions,
        278, 279, 281, 282
Translation lookaside buffer
    (TLB), 187, 311, 325
Tree insert operation, 204
Typed Object Identifiers
    (TOIDs), 92

U

Undo logging, 321
Uninterruptable power supply
    (UPS), 18
update_row() method, 270

V

Valgrind tools, 208, 209
Vector of strings, 171
Versioning, 193
vmemcache_add() function, 185
vmemcache_get() function, 185
Volatile libraries, 64
Volume managers
    advantages, 383
    Linux architecture, 384
    mdadm, 384
    NUMA systems, 383

W, X, Y, Z

Write pending queue (WPQ), 18
write_row() method, 268

Printed in the United States
By Bookmasters